THE

M000207477

IDEAS IN PROFILE
SMALL INTRODUCTIONS TO BIG TOPICS

ALSO BY FRANK CLOSE

An Introduction to Quarks and Partons

End: Cosmic Catastrophe and the Fate of the Universe

Too Hot to Handle: The Story of the Race for Cold Fusion

Lucifer's Legacy: The Meaning of Asymmetry

Particle Physics: A Very Short Introduction

Nuclear Physics: A Very Short Introduction

The Particle Odyssey: A Journey to the Heart of the Matter

*The New Cosmic Onion: Quarks and the
Nature of the Universe*

The Void

Neutrino

Antimatter

The Infinity Puzzle

*Half Life: The Divided Life of Bruno Pontecorvo,
Physicist or Spy*

Theories of Everything

FRANK CLOSE

PROFILE BOOKS

First published in Great Britain in 2017 by
PROFILE BOOKS LTD
3 Holford Yard
Bevin Way
London WC1X 9HD
www.profilebooks.com

Copyright © Frank Close 2017

10 9 8 7 6 5 4 3 2 1

The moral right of the author has been asserted.

All rights reserved. Without limiting the rights under copyright reserved above, no part of this publication may be reproduced, stored or introduced into a retrieval system, or transmitted, in any form or by any means (electronic, mechanical, photocopying, recording or otherwise), without the prior written permission of both the copyright holder and the publisher of this book.

All reasonable efforts have been made to obtain copyright permissions where required. Any omissions and errors of attribution are unintentional and will, if notified in writing to the publisher, be corrected in future printings.

A CIP catalogue record for this book is available from the British Library.

ISBN 978 1 78125 751 7

eISBN 978 1 78283 309 3

Designed by Jade Design *www.jadedesign.co.uk*

Index by Bill Johncocks

Printed and bound in Italy by L.E.G.O. S.p.A.

The paper this book is printed on is certified by the © 1996 Forest Stewardship Council A.C. (FSC). It is ancient-forest friendly. The printer holds FSC chain of custody SGS-COC-2061

Contents

1

Lord Kelvin's Hubris

In 1980, Stephen Hawking speculated that the end might be in sight for theoretical physics and that the arrival of a Theory of Everything might be imminent. Was he unconsciously echoing the assertion by the American scientist Albert Michelson in 1894 that 'the grand underlying principles have been firmly established. Further truths of physics are to be looked for in the sixth place of decimals',[1] or by Lord Kelvin in 1900 that 'there is nothing new to be discovered in physics now. All that remains is more and more precise measurement'[2]? Perhaps not.

Nature alone knows what extends beyond the horizons of our present vision, and it repeatedly reveals the limits of our imagination. Within a few years of Lord Kelvin's remarks, the discovery of the nuclear atom, and the rise of quantum mechanics and relativity, made the exuberance of those titans of nineteenth-century science appear naive. The truth, however, is more nuanced, and in consequence the implications are rather different. The words of Lord Kelvin (certainly) and Albert Michelson (to a degree) have been taken out of context and often misquoted. When carefully interpreted, what they actually said has a more profound message for seekers of the theory of everything.

Lord Kelvin's enduring and strongly held belief that the main role of physics was to measure known quantities to great precision had in fact inspired Michelson's remarks. Lord Kelvin had been impressed by Maxwell's theory of electromagnetic radiation as well as by thermodynamics, a description of heat based on mechanics, of which Kelvin himself was a prime architect. It might be possible, he felt, to understand the concept of energy in terms of the motion of particles, as the broad underlying principles seemed to be at hand.

On Friday 27 April, 1900, Kelvin gave a speech about his vision at the Royal Institution in London, the place where Michael Faraday had made the discoveries in the fields of electricity and magnetism that underpinned the new physics. Instead of making an uncritical claim that the synthesis of light, heat and mechanics meant that the end of physics was imminent, Kelvin began his speech thus: 'The beauty and clearness of the dynamical theory, which asserts heat and light to be modes of motion, is at present obscured by two clouds.'[3] This became known as the 'two clouds' speech.

In contrast to the folklore that Lord Kelvin was arrogantly announcing the end of physics, he was actually drawing attention to two outstanding puzzles. If he was wrong, it was in the hope that the 'two clouds' were mere puffs in an otherwise clear blue sky. In reality, they were the heralds of storms. Their removal would require the construction of two great pillars of twentieth-century physics: Einstein's theory of relativity, and the quantum theory.

So Lord Kelvin was wrong in detail, certainly, but he was nonetheless well aware of the limitations of late-nineteenth-century physics. Indeed, when he made those remarks, clues to the impending revolutions in twentieth-century physics were, with hindsight, already in plain view. This will be worth bearing in mind when we come to assess modern claims that the end of physics is once more in sight.

2

What is a theory of everything, and what is 'everything'?

Theories of everything can be roughly described as theories which draw on work in all relevant branches of current knowledge – physics, astronomy, mathematics, and so on – which seek to explain everything about the universe that is currently known. From this, it is easy to see that a theory of everything is a moving target. An explanatory account of the known universe may reign supreme for decades, even centuries. During that time it may be the basis for numerous scientific and technological advances. Then, perhaps as a direct or an indirect result of these advances, a new discovery is made, adding to the 'everything' which is known and which cannot be explained by the accepted theory in terms consistent with itself. A new theory of the new 'everything' is then required. And so the cycle continues.

Lord Kelvin's two clouds heralded paradigm shifts in our understanding of space and time, and of the microscopic structure of matter. Given that nuclear physics and quantum physics are so rich and far-reaching, and that Albert Einstein's relativity theory absorbed Isaac Newton's great works on mechanics and gravitation, one might wonder how nineteenth-century science could have been blind to them. The explanation of how these fundamental pillars of wisdom remained hidden for so long, while Isaac Newton, James Clerk Maxwell and Lord Kelvin created theories of everything then known, touches on profound properties of our universe, and arguably on our ability to successfully decode its laws.

A theory of everything (or TOE, as it is sometimes abbreviated) would have to describe nature across all distances, times and energies. Our experience is limited to a mere fraction of these vast ranges, though over the centuries it has grown. In practice, nature does not cover the spectrum homogenously, so we can build theories of subsets of phenomena where ignorance in one area need not prevent progress in others.

That we have been able to advance our understanding without having a theory of truly everything is a consequence of the way natural phenomena can be grouped into discrete regimes: they form what I have referred to as a 'cosmic onion', whose component layers are linked together but whose contents are, to an excellent approximation, independent of one another. A theory of everything-at-one-layer succeeds because nature effectively consigns manifestations of other layers to quarantine. Suitably isolated, they play no effective role in the description of phenomena at the layer of interest.

In this book I shall illustrate this compartmentalisation for the material universe in discrete scales of size, and quantify the different scales of energy, temperature or spatial resolution we need to study to expose their dynamics. For example, before the twentieth century, physics was limited to phenomena below the temperature of blast furnaces: the millions of degrees where nuclear physics takes over were out of range, let alone the thousands of trillions at which the Higgs boson bubbles into view.

Thus we can build a theory of everything, where 'everything' means 'within a specific limited range of energy'. That is how science has grown historically. It took centuries to reach the conditions revealed by the Large Hadron Collider at CERN, but along the way scientists developed a sequence of theories that were applicable to different ranges of energy.

For example, at the human scale such a theory already exists. Mathematical relationships accounting for everything bigger than the atomic nucleus have been with us since the work of Austrian physicist Erwin Schrödinger, the German physicist Werner Heisenberg and the Cambridge mathematician Paul Dirac, ninety years ago. The equations of this theory, which describe the behaviour of electrons and atoms, are taught to students. Their simplicity, however, is highly misleading as they are difficult to manipulate and impossible to solve except for a few simple cases. It is only with the development in recent years of powerful computers that the range of such solvable problems has grown. No one has deduced from these equations the properties of simple amino acids, let alone the workings of DNA, though this has hardly held back the astonishing development of modern biology. Similarly, starting from Isaac Newton's 'theory of everything-large-that-moves', we can predict solar or lunar eclipses with certainty, but not the weather.

Thus when Dirac's theory of everything is applied to the behaviour of electrons in the periphery of atoms, the complexities of the atomic nucleus can be isolated, and ignored.

The theory of everything-for-sequencing-the-genetic-code may flow from the symbols A, C, G and T, which represent adenine, cytosine, guanine and thymine – the linked units of nucleic acids of a DNA strand. Dirac's more fundamental theory of atomic physics and chemistry, which underpins the existence and structure of complex molecules, may be consigned to quarantine if your primary interest is the manipulation of those chains of amino acids encoded in A, C, G and T.

Even today, some energy domains have no theory at all, and the modern quest for theories of truly everything involves finding theories to cover the whole energy scale. The progress of science has not been restricted by the lack of an all-embracing theory of everything, nor by our inability to solve the equations of those 'theories of something' that have been formulated. One of the themes of this book is to consider whether the quest for a 'theory of truly everything' is a realistic goal, and to illustrate how practical science is largely independent of it.

The book's structure will illustrate how the gift of nature that enables science to quarantine areas of 'everything' has seeded the advance of theoretical physics down the centuries. Chapters 3 and 4 review this history up to the present day, starting with Newton's mechanics in the seventeenth century and its application to thermodynamics in the nineteenth. Electricity, magnetism and light were described by Maxwell's theory in the nineteenth century, but new data led to the birth of quantum theory and special relativity theory. The marriage of relativity, quantum theory and mechanics led to Dirac's fundamental theory, which underpins chemistry and the structure of DNA, and inspires the current standard model or core theory of particles and forces, with the recently discovered Higgs boson as its capstone. Chapter 5 describes theories of gravity, and their flowering in general relativity theory, while the challenge of finding a viable quantum theory of gravity is the theme of Chapter 6. In the concluding pages these ideas are brought together in assessment of the likely direction to a final theory of everything.

But first, the title of the book itself suggests two questions: what is a theory, and what is 'everything'? 'Life, the universe and everything' runs the mantra. In this book, 'life' and, to a large extent, 'the universe' will be in quarantine: 'everything' refers to the rest, namely the inanimate *contents* of the universe. The ultimate challenge for theoretical physics is to explain where those contents come from, understand the laws that govern their behaviour, and explain why they have the properties they do – properties that enable life as we know it to exist.

As for theories, creating them is all too easy. However, this does not mean that any theory is of use to science. Science is a body of knowledge based on demonstrable and reproducible data. If the data disagree with your theory, science demands that you revise the theory; this distinguishes science from cults, which reinterpret the facts to fit the theory. Experiment decides which theories describe nature, and which concepts are no more than beautiful ideas. In this respect, it is arguably easier to be Shakespeare or Bach than a theoretical physicist, for if a few words were altered in *Hamlet*, or a phrase modified in a Bach fugue, a work of art would remain, whereas to change a symbol in the equations of Einstein, or in the theory that leads to the Higgs boson, would cause the whole edifice to collapse. However beautiful a theory may be, if experiment disagrees with it, then the theory is scientifically redundant.

This is an appropriate moment to summarise the first requirement of a powerful theory, and certainly of one that would claim to be a theory of everything. A strong theory is one that pulls a range of disparate phenomena into a single concept, and which inspires new connections that can be tested against experiment. This requirement, that theory be amenable to experimental test (at least in principle), is the arbiter of what qualifies as science. This will form a coda to the questions of whether our universe is the one and only, what might have preceded it, and whether such questions lie within the realm of science.

3

Newton's theory of inanimate tortoises

Nature and Nature's laws lay hid in night.
God said, 'Let Newton be!' and all was light.

(Alexander Pope's epitaph for Isaac Newton,
who died on 21 March 1727)

In the seventeenth century, Isaac Newton produced his celebrated laws of motion. For the next two hundred years these were the accepted theory of everything for the dynamics of objects which are slow (relative to the speed of light) and large (relative to atoms). Newton showed how the mutual influence of one body on another results in changes in their motion. His formulation of what is now called classical mechanics consists of three laws, which are at first sight 'obvious' and deceptively simple.

The first of these laws is known as the law of inertia: a physical body will remain static, or continue to move at a steady constant speed, unless some external influence, a 'force', acts on it. The bigger the force, the greater the acceleration. Experience shows that if you apply the same amount of push to a tennis ball and to an identical volume of lead, the tennis ball will accelerate more than the lead: Newton decreed that the relative acceleration of the two bodies per unit force is a measure of their intrinsic inertia, or 'mass'. This is known as Newton's second law of motion.

Actually, the second law contains the first as a special case: if no force overall acts on a body, there is no acceleration and so the body continues on its way or remains at rest. The word 'overall' here could imply no force at all, as in Newton's first law, or there could be two or more forces that cancel one another out so that their net effect is zero. An example of the latter is your situation on Earth right now. The force of Earth's gravity pulls you downwards. That you are at rest relative to the vertical is because the floor is pressing up on your feet or your seat with an equal and opposite resistance. This state of affairs is often cited as an example of Newton's third law – to every action there is an equal and opposite reaction. It is the reaction to the downward pull of gravity that we experience as weight.

Newton's mechanics provided the theory of everything needed to describe the motion of slow, large objects. These restrictions on its universality were not appreciated, however, for another two hundred years. When Newton published his theory, in 1687, the speed of light had been measured just twelve years before. It was huge compared with the speeds of everyday experience. More than two hundred years would pass before Einstein showed that Newton's theory was inaccurate for bodies moving at speeds close to that of light. The atomic theory of matter was still a hundred years in the future, and the need for a 'quantum mechanics' to describe atomic dynamics was two centuries hence. Thus through the eighteenth century and for much of the nineteenth, for all practical purposes Newton's mechanics appeared to be the final theory.

CAN YOU SOLVE IT?

Having a theory is only a first step. To be useful in practice, it is necessary to compute its implications. Newton's laws predict the motion of snooker balls. When the cue ball hits the pack in a game of snooker, up to fifteen red balls may be set in motion as they collide, bounce and jiggle one another. Newton's laws can determine their trajectories, if you have the patience to carry out all the computations. For Victorian scientists, who were limited to algebraic calculations with slide rules as their most advanced tools, this was at the very limit of practicality. Today, computer programs can track the paths of particles. So some problems that were too complex to be solved in the past can now yield to computer codes. This is an example of how the development of new tools can advance the range of application of a theory of everything.

The power – and the limitations – of Newton's dynamics are apparent in their applications to the motions of the Moon and the planets, the rise and fall of tides and predictions of the weather. Isaac Newton published his theory of dynamics in his book *Principia* in 1687. He presented his laws of motion, and used *gravitas* – the Latin word for weight – in his law of universal gravitation. This famously arose from his insight that falling apples and the motions of planets are all governed by gravity. Unlike the electrical attractions and repulsions resulting from positive and negative charges within atoms, which cancel one another out, the individual gravitational attractions exerted by each and every particle in a large body add up. The Sun, no bigger than a thumbnail when viewed from the Earth, traps the planets in a cosmic waltz across hundreds of millions of kilometres of space. Newton posited that gravity's pull between two bodies diminishes as the square of the distance between them increases, and that a massive body such as the Sun extends its gravitational tentacles into space uniformly in all directions.

The Earth's orbit around the Sun is an ellipse, but near enough to a circle not to affect the following thought experiment. Picture the Sun at the centre of a sphere whose radius is the same as that of the Earth's orbit. The gravitational tug on our planet is the same at all points on the surface of the sphere. If we now imagine ourselves transported to an orbit that is twice the radius of the Earth's, the surface of the corresponding sphere will be four times greater, as the area increases with the square of the distance. Newton realised that if the force of gravity were likened to tentacles spreading out from the source in all directions symmetrically, then the intensity of the force at any distance would be spread uniformly across the area of the imaginary sphere. As the area of the sphere increases with the radial distance squared, the intensity at any point on it weakens. This model rationalised the inverse square law of gravitational force.

These analogies highlight the intimate relation between the behaviour of these forces and the three-dimensional nature of space. It rationalised Newton's hypothesis that the force of gravity acts instantaneously between two apparently disconnected bodies. The region of space around a body is filled with its gravitational field, and the body exerts a force on any objects within that field. It is the Earth's gravitational field stretching into space that pulls skydivers to ground, and the Sun's gravitational field that keeps the Earth in its annual orbit.

Newton's theory explains why each planet orbits the Sun in an ellipse, with the Sun at one focus of the ellipse. It also predicts that the speed of a planet varies as the inverse square root of its distance from the central Sun. Thus Saturn, which is slightly more than nine times further from the Sun than we are, moves at about one-third of our speed, and takes nearly thirty times as long to complete a circuit. Newton's was a clockwork universe, where planets follow permanent, regular orbits; the design seemed to accord with the perfection expected from a divine creator. But this ideal would not last.

The Moon has completed more than 20,000 orbits of the Earth since the time of the Egyptian philosopher Ptolemy, in the second century AD. Near the end of the seventeenth century, the Oxford physicist and then Astronomer Royal, Edmond Halley, examined records of medieval and ancient solar eclipses over that timescale. He discovered that when he used the position and trajectory of the Moon to determine retrospectively when solar eclipses should have occurred, the times he calculated differed from the actual ones by up to an hour. Halley deduced that in the past, the Moon must have moved across the sky from east to west more slowly than in his own time.

This was a far-reaching, even heretical assertion. For the Moon to have changed its motion in such a way would imply that its course through the heavens did not repeat in regular orbits. Such changes in its orbit could eventually cause the system itself to disappear, with the Moon falling into the Earth or escaping into space. For many philosophers, to theorise that the cosmos could decay in this way was a slur on the Almighty, as it implied that God was such an unskilled craftsman that He had constructed a system of stars and planets that could fall into ruin and disorder. Nonetheless, Halley was right, as even the fundamentalists were eventually forced to concede. The question now became: what causes the cumulative acceleration of the Moon?

There are two contributions, one of which, discovered by the French mathematician Pierre-Simon Laplace in 1776, exposed a fundamental limitation to Newton's theory. Laplace demonstrated that orbits would eventually degrade if, in contrast to Newton's theory of instantaneous action at a distance, gravitational forces take time to be transmitted across space. This concept would later become a plank of Einstein's theory of gravity. Laplace's calculations accounted for about half of the observed effect; the second contribution, which takes care of the rest, illustrates some of the complex subtlety in Newton's theory.

The Moon raises tides on the oceans. As the Earth rotates, the Moon drags these tidal bulges with it, so if the Moon is directly overhead, a high tide will soon follow. (The Moon overhead is followed by a high tide, but not every high tide happens when the Moon is overhead. There are two high tides each day, half of which occur when the Moon is on the far side of the Earth.) This bulge in turn exerts a gravitational pull on the Moon, slowing its rotation and also the Earth's: days are gradually growing longer, as are the lunar months, though the number of days per month is falling. In the far future there will be only one day per month, the Earth always presenting the same face to the Moon, as the Moon does to us today.

Modern technology confirms these theoretical calculations. Atomic clocks measure the length of the year to fractions of a microsecond – the midnight hour is periodically adjusted – and laser range-finding using a mirror placed on the Moon by Apollo astronauts confirms that the Moon is gradually moving away from us. There is no permanent clockwork cosmos. Newton's theory ran contrary to fundamentalist prejudices, but its implication of an inconstant cosmos proved consistent with reality.

Newton's theory of gravity describes perfectly the motion of the Earth around the Sun, and of the Moon around the Earth. This poses another question: if we specify the state of all three, namely the masses, positions and velocities of the Earth, Moon and Sun at some instant, can we determine their future trajectories just by applying Newton's laws of motion? This is known as the three-body problem.

In 1887, the French mathematician Henri Poincaré showed that there is no general algebraic solution for such a situation. In special cases, such as when the three bodies form an equilateral triangle (which for the Sun–Earth–Moon system cannot happen in practice), algebraic solutions do exist, but for an arbitrary configuration three bodies is one too many. One reason for this is that the motion of three mutually gravitating bodies does not in general repeat so as to form a closed curve whose shape would be describable by some analytic formula.

The technique for solving the problem is to do so iteratively – that is, by making a series of approximations that lead to ever more accurate representations of the true answer. First, treat the Moon and the Earth as a single entity which orbits the Sun. Then you calculate the motion of the Moon around the Earth, meanwhile neglecting their combined motion around the Sun. The answer can then be refined by considering the effect of the Sun on the Moon's motion around the Earth as a perturbation.

Here we see an instructive example of a limitation of a theory of everything. We may have determined the equations, but this does not necessarily mean that we can solve them exactly. For three objects with widely different masses and separations, as in the Sun–Earth–Moon system, a practical solution to Newton's equations can be found by sequential perturbations. For a large number of objects, however, this method fails. This is a general constraint, one that applies not only to objects that interact by the gravitational force. The mutual electrical interactions among large numbers of atoms, or the turbulent motions within the atmosphere, cannot be solved exactly. Instead, models have to be developed which are based upon the underlying theory. We shall come back to this later when we meet thermodynamics and find that familiar concepts such as temperature and pressure are interpreted as measures of the states of motion and dynamics of a macroscopic system's constituent particles.

NEWTON'S UNIVERSE

The inverse square law of gravity's weakening with distance is critical for the structure of the universe – and also possibly for the development of physical science. The Sun contains 99.8 per cent of the total mass of the solar system, and Jupiter accounts for most of the rest. Jupiter is some three hundred times more massive than Earth, yet it is so remote that, thanks to the inverse-square enfeeblement of force with distance, it exerts no noticeable gravitational pull on the Earth. Contrast this with the small but relatively nearby Moon. The Moon exerts a gravitational tug that causes the tides, but neither the other planets nor remote galaxies of stars have any measurable effect on the oceans.

Tides, eclipses and the orbits of artificial satellites can be determined without needing to take account of those distant masses. If the force of gravity tailed off in direct proportion to distance, rather than with the inverse square of distance, the gravitational tugs from the massive outer planets – Jupiter, Saturn, Uranus and Neptune – would, on the Earth, rival that of the Moon. In such an artificial universe it is possible that we could inhabit a planetary Earth, but unlikely that we could determine the rules of gravity: the ability to ignore all but two bodies, with small perturbations from a third, is what has enabled computations to be made and the basic rules formulated. In a universe where the distribution of stars is roughly uniform, their number will grow in proportion to the square of their distance. The amount of mass would then grow with the square of the distance and thus faster than the depletion of its gravitational influence, which would tail off in direct proportion to distance. The gravitational force of these remote stars would dominate those of the Sun and the massive planets.

In the real universe, where gravity's pull decreases as the inverse square of distance, the remote galaxies generate a constant background of gravity, though it is feeble compared with the pulls of the Sun and Moon. This gravitational shell does make its presence felt, however. To experience it, take a ride on a roundabout. To appreciate the sensation of being on a roundabout, and how this eventually helped inspire Einstein to improve on Newton's theory of gravity, first we need to examine the terms and conditions that apply.

To start with, we need to re-examine the basics. Motion means that the position of an object at one instant differs from its position at another. A lawyer might ask, 'What defines position?' to which a reasonable answer is, 'Relative to me.' In general, the position or motion of a particle can be defined only relative to some frame of reference. Newton envisaged some absolute space and time, a metaphorical grid of invisible measuring rods defining up–down, left–right and front–back: the three dimensions of space. Bodies which, relative to this matrix, are at rest or in uniform motion (i.e. not accelerating) move according to Newton's laws of motion. This grid forms the mental construct of what is known as an inertial frame.

In Newton's theory, any two inertial frames must have their associated grids of rods moving relative to one another at a constant speed (which could be zero), in a straight line and without rotation. Clocks in the two frames always show either the same time or differ from one another by a constant fixed amount. Thus Big Ben in London and the clock in New York's Grand Central Station show times that differ by five hours, due to the convention of time zones, but intervals of time are the same at both locations. If two events occur simultaneously in one inertial frame, they will do so in another. The matrix remains the same as we move through it. The tick-tock of Newton's metronome also doesn't change.

There is no absolute state of rest; only relative motions are unambiguous. Contrast this with acceleration, however, which has the same magnitude in all inertial frames. A commercial for a sports car which states that it 'accelerates from rest to 60 miles per hour in three seconds' needs no 'as viewed by a stationary bystander' caveat. Everyone agrees on this change in speed per unit time, or acceleration.

Passengers in this accelerating sports car will see their surroundings flash by increasingly rapidly. They will also feel themselves pressed into their seats, as if by some unseen force. There is no doubt that it is they and not the surroundings that are accelerating. As they hurtle round a corner, the passengers will feel themselves thrust to one side by what we call centrifugal force. During the car's acceleration, the passengers could deduce, at least roughly, the magnitude of the force they experienced. The car and its contents are not in an inertial frame.

And now we come to the experience of being on a roundabout, which provides another demonstration of the effects of acceleration. Imagine that you are on this roundabout and blindfolded. Isolated from the surrounding universe, you can nonetheless tell that you are rotating relative to … something. Newton's matrix of measuring rods has no visible presence, and there are no material objects to show you that it is there, but as you rotate you can feel it passing through your being. It is this effect that you feel as your orientation changes that we call centrifugal force.

Two hundred years after Newton, the Austrian physicist and philosopher Ernst Mach proposed that it is the distant stars that give rise to our sense of absolute relativity. This is the sensual consequence of gravity, governed by Newton's inverse-square law, permeating all of a space in a universe populated with stars throughout. Their numbers grow as the square of their distance from us, and their individual gravitational tugs decrease as the inverse square of the distance. The net effect is constant.

In a static universe, the sum of the gravitational pull of all the stars would be infinite. We now know that the universe is expanding, but the sense of gravity's web nonetheless remains. Einstein would build on this and extend Newton's theory of gravity, but not for another two hundred years. Until the early twentieth century, Newton's laws of motion and theory of gravity were the theory of everything mechanical in a clockwork universe.

It did not last: the devil, shouting, 'Ho.
Let Einstein be,' restored the status quo.

(J. C. Squire, 'In Continuation of Pope on Newton')

PERPETUAL MOTION

Newton's laws of motion contain an enigma: they do not distinguish between forwards or backwards in time. According to his mechanics, if you reverse a movie of any sequence of events, such as the paths followed by bodies that collide or interact with one another, the result is a sequence that is still consistent with his laws of motion. For example, consider a game of snooker which has reached the stage where only the black ball and the white cue ball are on the table. It is possible for a player to play a 'stun' shot, where the white cue ball strikes the black, stops, and transfers all of its momentum to the black ball. A film of this played in reverse would show the black ball hitting the stationary white one and coming to rest as the white ball recoils. Each of these is allowed by Newton, and consistent with our experience.

Now contrast this with the start of the game. The fifteen red balls are arranged in an ordered triangle, then the cue ball strikes the pack and disturbs them. As we saw earlier, analysis of their motion is complicated but in principle possible; everything agrees with Newton's mechanics. But when the cue ball strikes the pack, there are innumerable different ways in which the reds can split from the pack, which makes every game of snooker unique after this opening shot. Play this in reverse, however, and we would see a sequence that is almost certainly beyond our experience: up to fifteen red balls are moving in a variety of directions, miraculously jostling until they come to rest in a neat triangle, and then imparting all of their momentum to the white ball, which rushes down the table and is brought to a sudden halt by the tip of a cue.

This situation is not impossible. If fifteen red balls spread out on a snooker table could be set in motion, with each given precisely the right momentum, the result would be as just described. Newton's laws allow this to happen, but it is highly unlikely to do so in practice. If you saw this sequence in a movie, your natural reaction would be that the film was being run backwards. The moral is that, although there is no preferred arrow of time (i.e. heading either forwards or backwards) in Newton's fundamental theory, a natural sense of time emerges when a large enough number of particles are involved. What is illustrative for fifteen snooker balls is even more evident for an assembly of trillions of atoms. Eggs break when they fall to the ground; shattered pieces do not rejoin to make a perfect egg.

This illustrates how phenomena which are not apparent or even present in the fundamental theory can emerge from complexity when large numbers of atoms act cooperatively. The phenomenon of the arrow of time was known even before Newton: after all, things decay and people grow old. The first serious scientific theories of this vital asymmetry were born in the nineteenth century with the Industrial Revolution and the emergence of thermodynamics. This historical development shows how a successful practical theory of macroscopic systems can be constructed, even though the fundamental underlying laws have not been identified. Later, as links with these more basic concepts are made, new insights emerge.

The Scottish engineer James Watt built the first effective steam engine in 1782. The basic idea was to burn coal, which would heat water to make steam, and then use the pressure of the steam to drive a piston or turn the blade of a turbine. To do this efficiently, the physics underlying the entire process needed to be understood. So was born the science of thermodynamics: the motion of heat.

Today we visualise this in terms of the motion of molecules in the water or steam, but such concepts were foreign to nineteenth-century scientists. They identified macroscopic properties, such as the pressure in the piston, the volumes of liquid and gas, and – of fundamental significance – temperature. Rules were determined which accounted for the behaviours and interrelationships of these properties.

For example, the Irish scientist Robert Boyle, working in Oxford in the seventeenth century, had shown that, at a fixed temperature, the application of pressure to a gas compresses its volume in proportion to the pressure. Newton himself showed that this could be explained if the gas consisted of impenetrable small particles which were at rest and repelled one another with a force that is proportional to their separation. Here, Newton had a tentative theory of Boyle's observation that the product of pressure and volume remains constant. However, it was hardly an explanation. Newton supposed the force was proportional to the distance of separation in order to account for the observed behaviour, but he had nothing to back his supposition. His theory worked, nonetheless, so long as the pressure did not become too great and the temperature remained constant.

These two shortcomings highlight the limitation of Newton's theory. First, it is an approximate description. Although the volume of a gas appears to reduce in direct proportion to the applied pressure, this is only true as long as the pressure is not too great. Invoking the 'sixth place of decimals' mantra, precise data would have revealed that the volume actually decreases very slightly more slowly with pressure, an effect which becomes more noticeable only at high pressures. As to why this happens, a simple picture emerges when we examine the central assumption of Newton's theory: he had supposed the fluid to consist of dimensionless point particles, which therefore cannot overlap. In reality, particles of non-zero size are squeezed into one another and forced to compete for space. This creates a resistance to further squeezing, and the volume decreases less rapidly.

The second limitation is more obvious. Newton's theory was fine as an interpretation of Boyle's law, which is for a gas at constant temperature, but it was not a theory of everything about gases. For example, it is possible to reduce volume not only by applying pressure, as Newton recognised, but also by cooling the gas. Newton's theory took no account of temperature, nor did it give any explanation of what it actually is. Temperature entered the description in 1738, when the Swiss mathematician Daniel Bernoulli added motion to Newton's static picture. He postulated that air consists of 'very minute corpuscles, which are driven hither and thither with a very rapid motion'.[1] The greater their mean energy, the higher the temperature. He explained that the force exerted by these particles hitting the edge of a container is greater at a higher temperature. Furthermore, for particles in a liquid this motion would tend to push the boundaries of the fluid outwards. In other words, the volume would expand with temperature.

Here we have the kernel of the modern picture of temperature as the macroscopic measure of the kinetic energy of constituent particles. However, this picture became accepted only a century later, thanks to James Joule, a Manchester physicist and brewer. Joule wanted to operate his brewery as economically as possible. His experiments to determine the most efficient source of power led to his demonstration of the quantitative equivalence of heat and work.

In simple mechanical terms, work is a measure of how much force has been applied to something, and over what distance. Work is the product of these two measures, and when work is done on a body it gives that body energy of motion – kinetic energy. Joule showed that whatever the form of work – whether, say, a push by a piston, induced by an electric motor (of which more later) or the result of a fall under the force of gravity – a given amount of work would always generate the same amount of heat. In recognition of this, Joule's name has been adopted as the unit of energy, the joule (with a lower-case j). The name of Watt is the unit of power, one watt being the rate of energy transfer measured in joules per second.

Here we see the impact of the Industrial Revolution on the development of science and the understanding of energy. The equivalence of work and heat as forms of energy is the basis of the first law of thermodynamics: energy can be changed from one form to another, but is always conserved overall.

Although Joule had shown that heat and work are equivalent and interchangeable in the energy accounts, they are not symmetric. Any form of work can be turned entirely into heat, at least in principle, but the converse is not true, as heat is dissipated when converted back to work. Not all of the heat in a steam engine will drive the piston: some is lost as heat to the surroundings, or as friction in the bearings. The total amount of energy is conserved, in line with the first law, but some of it remains as heat instead of being converted into useful work.

The German physicist Rudolf Clausius was first to recognise that heat loss is irreversible. His insight, in 1850, is at the root of our perception of the arrow of time: the 'nature of nature' is that it is one-way. This led William Thomson, later Lord Kelvin, to declare the second law of thermodynamics: mechanical work inevitably tends to degrade into heat, but not vice versa. As we shall see later, 1865 was a remarkable year for theories of everything. The first breakthrough that year was when Clausius gave the second law of thermodynamics a quantitative description: he introduced the concept of entropy.

The word 'entropy' comes from the Greek *en* for 'in' and *trope* for 'turning'. Clausius focused on the transformation associated with any process. Entropy, he said, would remain constant if a process is reversible, but would change for an irreversible process. For any system which is isolated from the rest of the universe, entropy can never decrease. The dissipation of heat which is lost to useful work is associated with an increase in entropy; the forward march of time, as perceived in macroscopic phenomena, is also a manifestation of entropy's increase.

Entropy is thus a measure of the state of a macroscopic system, as are pressure, volume and temperature. The concept of entropy may be less intuitively familiar than the latter trio, but they all share the property of being quantitative measures which describe the state of a system. Changes in their values, and correlations among them, are useful in describing the dynamics or in accounting for the efficiency of a steam engine, or of any engine driven by heat, for example. For the nineteenth-century physicist or engineer, these concepts were the elements of the theory of everything concerned with thermodynamics. Lord Kelvin, who had played a seminal role in establishing the science of thermodynamics, realised that these properties could all be understood in terms of Newton's theory of motion – surely justification for his proclamation that nothing in physics remained to be discovered.

THE ARROW OF TIME

We have seen how Newton's laws contain no sense of time's arrow, but when they are applied to a system of many connected particles, a sense of forward and backward, allowed and forbidden, emerges. Entropy is the link between the second law of thermodynamics, with its sense of irreversibility and an arrow of time, and the emergence of that arrow from Newton's laws when many particles are involved. Entropy is a measure of order and disorder. The arrow of time proceeds in the direction of order turning into disorder, which is equivalent to entropy increasing in any system that is isolated from the external world.

An isolated system can evolve until its entropy is at its

maximum possible magnitude. At that point, the system has arrived at its most random state. There is then no possibility of further change: the system has reached what is called thermodynamic equilibrium. At the other extreme lies a hypothetical state of total order and minimum entropy. The third law of thermodynamics is that this state can be attained only at the absolute zero of temperature, minus 273 degrees Celsius.

The ability of a refrigerator to cool its contents and decrease their entropy appears at first sight to run counter to the second law. However, a working fridge is not isolated from its surroundings: it receives energy from the electricity supply. The interior of the fridge may become cold, but the effect overall is to dissipate heat into the surroundings. If you have any doubt, feel the warm pipes at the rear of a fridge.

The underlying theory is revealed if we consider that temperature and heat are both manifestations of the kinetic energy of constituent particles. The motion of the particles manifests as the temperature; their mutual arrangement is the root of entropy. Some familiar examples will illustrate this.

The snooker game has the essential ingredients. There is only one pattern in which the pack of fifteen red balls can be arranged at the start of the game; this is a state of high order, or low entropy. Contrast this with the situation after the cue ball strikes them: there are vast numbers of possible outcomes; the entropy is now high. The increase in entropy, or disorder, is in accord with the second law of thermodynamics and with our sense of past and future.

Another common misconception is that the emergence

of life on Earth, with order arising from disorder, runs counter to the second law. Some Creationists use this to claim that life could not have emerged other than by some divine intervention, and that the latter took place just 6,000 years ago. This is actually a theory of nothing, as I shall now argue.

The resolution of this apparent paradox is that the Earth is not isolated, nor is it in an environment of thermodynamic equilibrium. The sky contains a hot Sun in a cold background – a prime example of a non-equilibrium situation. For every high-energy photon that arrives here from the Sun, the Earth radiates typically twenty lower-energy photons into the cold surroundings of space. This is in accord with the conservation of energy, the first law of thermodynamics. The second law is also satisfied: the number of possible ways in which the total energy could be shared among this score of photons is vast, so an initial state of relatively high order – a single photon – has led to a disordered situation. The increase in entropy associated with the radiation of all these low-energy photons into space exceeds the local decrease on Earth by which the biosphere operates.

Starting from utter disorder, the entire biomass of the Earth could be converted into a state of high order by such processes. And how long would that take? As far as the second law is concerned, it turns out that about a year would be enough. It is thus ironic that the Creationists have aimed at the wrong target: physics, far from being inconsistent with human existence, would allow the entire biomass to emerge within a single year, certainly within 6,000. Over to the biologists as to why it actually took billions!

Today we know that the universe, at least what we can see of it, began in a hot Big Bang some 13.8 billion years ago. If the universe is indeed a closed system, its entropy should be increasing. Here we have a new conundrum, whose resolution is still debated: how did the initial state of low entropy arise? The second law of thermodynamics has turned out to be key for understanding not just the behaviour of our universe, but also its very existence.

In the late nineteenth century, this conundrum was unknown. Newton's mechanics appeared to be consistent as a theory of everything to do with motion, and thus at the root of a kinetic theory of heat. Although we cannot solve its equations for the cases of trillions of atoms in a solid or fluid, we do have a qualitative understanding of how macroscopic phenomena such as temperature and entropy emerge, enabling us to construct practical effective theories of macroscopic thermodynamics. Newton's mechanics have also explained how the arrow of time emerges. No wonder Lord Kelvin expressed such satisfaction.

The link between temperature, heat, work and energy is at the root of our historical ability to construct theories of matter, such as Newton's dynamics, while ignoring – and indeed, being ignorant of – atomic dimensions. The kinetic energies of particles in a fluid at room temperature are typically about 1/40 of an electronvolt,[2] where one electronvolt (1 eV) is shorthand for an energy of a mere 1.6×10^{-19} joules. The joule is an appropriate unit for macroscopic physics, but it is no more sensible to use it for the energies of individual atoms or molecules than it would be to describe the width of a human hair as a fraction of a mile. It would not be wrong, just inefficient. In the white heat of the Industrial Revolution, atoms reached energies of about 1 eV. We shall see later that, to resolve the innards of atoms, the energies required are billions of electronvolts, far beyond what was accessible before the twentieth century. This is a reason why Newton's laws and thermodynamics proved to be practical and useful theories in the nineteenth century without any need for theories of atomic physics. The second law of thermodynamics emerged, as we saw, from the insights of Clausius in 1865. That same year, James Clerk Maxwell produced his theory of everything to do with electricity, magnetism and light. As we shall see, this theory would expose limitations in Newton's mechanics and inspire Albert Einstein to construct the foundations for modern theories.

LET THERE BE LIGHT!

By the middle of the nineteenth century, a substantial range of phenomena had been incorporated into the realm of mechanics. Newton had defined its basic laws, which described the dynamics of heavenly bodies and the operation of machines on Earth, and Joule had determined the exchange rate between work and heat, which led to thermodynamics, and a dynamical theory of heat. These did not constitute 'everything', however, because in addition there was a whole continent of other apparently disparate phenomena, which could be lumped under the headings of electricity and magnetism, and also optics – the nature of light.

Electric and magnetic phenomena had been known for centuries. By the nineteenth century, the prevalent ideas were that electric and magnetic fields mediate the forces between electrically charged particles or magnets by analogy with the gravitational field and Newton's mechanics. Unlike gravity, which is always attractive (there is no 'antigravity'), electric charges come in two varieties, labelled positive and negative. The rule 'like charges repel, unlike charges attract' described their behaviour. The properties of electric and magnetic fields were encoded in four basic laws which provided the foundations for the construction, in 1865, of the theory of everything then known about them.

Carl Friedrich Gauss was a German mathematician who, in the early years of the nineteenth century, contributed to many areas of science. His work led to the eponymous Gauss's law, which describes electric fields and electric forces. Essentially, it states that the flux of an electric field leaving a volume is proportional to the amount of electric charge within that volume. One consequence of Gauss's law is that the force between two electric charges diminishes like the force of gravity: in proportion to the square of the distance between them.

The second of the laws was Gauss's law for magnets. This stated that there is no net flux of magnetic field out of any volume: what flows out is balanced by what flows in. A consequence of this law is that there can be no isolated sources of magnetic charge. Bar magnets, for example, are bipolar, with a north pole matching a south.

The Earth is itself a vast magnet, with north and south magnetic poles, and is surrounded by a magnetic field. The source of this field is a swirling current of electrical charge deep within the planet's core. Here we see a profound connection between electricity and magnetism. The electrical generation of magnetic fields is encoded in Ampère's law, named for the French physicist André-Marie Ampère. This, the third of the quartet, states that if an electric current passes through a surface, such as this page, a magnetic field will circulate on that surface, with a strength proportional to the magnitude of the current.

The final member of the quartet is Faraday's law. Michael Faraday made so many discoveries at London's Royal Institution that, had Nobel Prizes existed in the nineteenth century, he might well have won several. One thing he demonstrated was that it is possible for magnets to create electric fields. This was complementary to Ampère, who had focused on electric currents as sources of magnetic fields. Faraday observed that steady motion is not enough, however: the magnet first has to be accelerated, either by being moved suddenly or rotated. This is known as induction, and is the principle behind electric generators.

The ability of electric and magnetic fields to ebb and flow is harnessed in the electric motor, which helped to power the Industrial Revolution. The underlying theory, such as it was, relied on a set of rules, of which the above quartet was the most powerful, but they were still something of a hotchpotch. It was the Scottish physicist James Clerk Maxwell who created the theory of everything electromagnetic, in 1865. He united electricity and magnetism into a description of electromagnetism, and in the process showed that light is an undulating wave of electric and magnetic fields.

Furthermore, his insights into the relations between electric and magnetic fields led Einstein to expose limitations in Newton's theory of mechanics.

The seeds of relativity become apparent with Ampère's demonstration, at the start of the nineteenth century, that electric currents generate magnetic fields. An electric current consists of electric charges in motion, which prompts the question, 'In motion relative to what?' We saw in the previous chapter that a similar question arose with regard to position, and similarly a reasonable answer would be, 'Relative to you in your (static) inertial frame.' However, suppose now that you moved alongside the wire carrying the current, and at the same speed as the flow of electric charges inside it. You would perceive the charges to be at rest. Gauss's law tells us that an electric charge at rest, in an inertial frame, gives rise to an electric field, so you would perceive there to be an electric field whereas previously you perceived magnetism.

The implication is that when you alter your speed, electric fields emerge at the expense of magnetic fields, and vice versa. This illustrates that electric and magnetic phenomena are profoundly intertwined; what you interpret as electric or magnetic thus depends on your own motion. The identification of a field as electric, or as magnetic, may vary from one observer to another, but the results of experiments do not. Electric and magnetic descriptions are like two languages: the transcription may vary, but the message does not, provided the rules of translation are correctly applied. Maxwell's achievement was to construct a theory that includes this translation automatically. In so doing, he united electric and magnetic fields into what we call the electromagnetic field. Among the profound consequences are that Newton's mechanics could be at best an approximation to a richer description of the natural world.

Maxwell's genius was his perception that the laws of electric and magnetic fields should have a mathematical similarity. Faraday's law explained how electric fields arise if a magnet jiggles – technically, if a magnetic field changes with time. Contrast this with Ampère's law for the generation of magnetic fields from electrical effects, in which there is no mention of time dependence: a steady electric current was sufficient. Maxwell realised that this description was incomplete. When an electric current is interrupted by a capacitor, an electrical component in which positive and negative charges build up on two separated plates which have been connected to an electricity supply, an electric field builds up between the plates. Maxwell extended Ampère's law to include this time-dependent electric field. He now had encoded the four laws in a quartet of mathematical equations, which were profoundly linked (see Figure 1).

In this form, with his extension of Ampère's law, Maxwell's equations summarised that a change in either an electric or a magnetic field would generate its complementary partner: electric to magnetic, and vice versa. An electric field is what is called a vector field, one that has not just a magnitude but also a direction – is a charge attracted or is it repelled, for example? One of Maxwell's equations implies that if the electric field oscillates such that the directions 'uphill' and 'downhill' are interchanged a certain number of times each

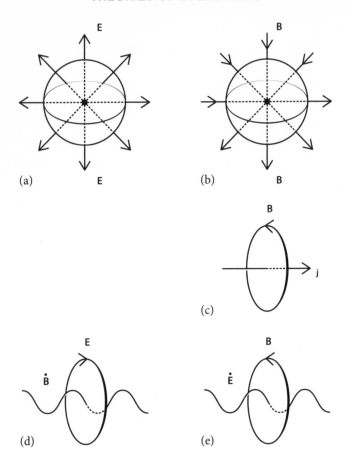

Figure 1: Pictorial summary of Maxwell's equations. The four basic elements of Maxwell's equations are illustrated here diagrammatically: (a) Gauss's electric law, (b) Gauss's magnetic law, (c) Ampère's law, (d) Faraday's law and (e) Maxwell's extension of Ampère's law. **E** and **B** denote, respectively, electric and magnetic fields. **Ė** and **Ḃ** (with a dot superimposed) denote electric and magnetic fields that change with time. The **j** denotes electric current.

second, the resulting magnetic field will also oscillate at the same frequency.

Another of Maxwell's equations symmetrically predicts that this oscillating magnetic field produces a pulsating electric field. Put this electric oscillation back into the original equation, and you find that the changeover goes on and on: electric to magnetic, back and forth. The equations predicted that a whole melange of electric and magnetic fields would spread across space as a wave.

Faraday's measurements of electric and magnetic phenomena provided the essential data that, in effect, show how much resistance empty space offers to the rise and fall of electric and magnetic fields. When these data are inserted into Maxwell's equations, they predict the speed of the waves to be about 300,000 kilometres each second, independent of frequency. This remarkably high speed happens also to be the speed of light. Maxwell now made his inspired and seminal leap: light is an electromagnetic wave!

The rainbow of colours we see with our eyes consists of electromagnetic waves whose component electric and magnetic fields oscillate hundreds of millions of times each second, the distance between successive crests in intensity being in a narrow range slightly below a millionth of a metre. Maxwell's insight implied that there have to be other electromagnetic waves beyond the rainbow, travelling at the same speed as light but oscillating with different frequencies. Here we see how a viable scientific theory has predictive power – something that elevates it above being simply a beautiful construct. A theory is powerful when it both explains known facts and makes new and testable predictions.

Infrared and ultraviolet rays were already known, the 'infra' and 'ultra' referring to the frequency relative to those of visible light. Maxwell was thus unifying a range of previously disparate phenomena. These clues inspired scientists to look for other examples. The German physicist Heinrich Hertz produced electric sparks and showed that they sent electromagnetic waves across space without the need for material conductors (this was the original source of the name 'wireless'). These primitive radio waves are electromagnetic waves akin to light but in a different part of the electromagnetic spectrum. They travel through empty space like rays of visible light, and at the same universal speed of 300,000 kilometres per second.

Maxwell's theory implies that electrically charged bodies and magnets interact with one another by means of the electromagnetic field, which spreads out from one body to another at the speed of light. Thus an electromagnetic force does not act instantaneously. Jiggle an electric charge at one location, and it is only when the resulting electromagnetic wave reaches a remote charge that the latter will start oscillating in concert. This is utterly different to Newton's mechanical picture, where such action occurs instantaneously.

SPACE, TIME AND SPACE-TIME

In Newton's picture, light in effect travels at infinite speed. In reality, as Maxwell's equations predict – and as experiment has demonstrated – the speed of light is exceedingly fast, but finite nonetheless. This was the first hint that Newton's theory might be just an approximation to a richer theory, and in practice valid only for situations where the speeds of bodies are trifling relative to that of light. As every situation that science had encountered until the nineteenth century satisfied this condition, Newton's mechanics was excellent in practice. Any deviations would lie in the sixth place of decimals, or be exposed unavoidably when exceedingly high speeds were involved.

The clues to the richer theory of mechanics are contained in Maxwell's equations for electromagnetism and light. Albert Einstein would build on Maxwell and subsume Newton's mechanics into his special relativity theory. Here, 'special' means restricted to a universe in which we can ignore gravity; his later extension of relativity to include gravity is known as his general theory. These theories are essential foundations for a final modern theory of everything.

Einstein's basic axiom was that the laws of physics do not depend on the uniform motion of an observer. Maxwell's theory of electromagnetism implicitly respects that: what one observer sees as an electric field, another might perceive as magnetic, but when these are united into electromagnetism, à la Maxwell, both observers will agree on the results of experiments.

The inspiration for Einstein's special relativity theory came when he started to think about how observers in different inertial frames would experience electromagnetic phenomena. Indeed, his famous paper was titled 'On the Electrodynamics of Moving Bodies'. This was the maturation of musings that had begun when, at the age of 16, he wondered what it would be like to ride on a light beam. This led him to a paradox, whose resolution would expose limitations to Newton's mechanics.

If you were travelling at the speed of light towards a mirror, and used Newton's theory of motion to predict the consequences, what would you expect to see in the mirror? Not your image: according to Newton, light is heading towards the mirror at the same speed as you, and will arrive at the mirror only when you crash into it. Thus the mirror is as yet unaware of your existence, and cannot reflect your image.

This is psychologically weird, and in fact no such thing would happen because there is a physical inconsistency with what Maxwell's theory would allow. In Newton's mechanics, if you travel at the speed of light alongside an oscillating wave of electric and magnetic fields, you would perceive an electromagnetic field that is oscillating in space from side to side, but not moving forwards. Yet there is no such thing in Maxwell's equations: oscillating electromagnetic fields move at a uniform speed of 300,000 kilometres per second. Thus if Maxwell's theory of electromagnetic phenomena is correct, it must be impossible for you to travel at the speed of light.

This contradicts Newton's description of motion. According to him, in principle there is no such thing as a universal speed limit. If I see you in a car coming towards me at 20 metres per second from the west, and there is another car coming towards me at 10 metres per second, but from the east, we can be sure that you will perceive this second car to be heading towards you at 30 metres per second. If these speeds were 200,000 and 100,000 kilometres per second, the same logic implies that you would perceive the relative speed to be 300,000 kilometres per second. This is identical to the speed of light, but if it is impossible to travel at the speed of light, relative to anything, this 'obvious' rule for the addition of velocities, which is inherent in Newton's mechanics, must somehow fail.

Because Newton's rules work so well in our everyday experience, Einstein deduced that they must be only an approximation to a more complete theory. Newton's dynamics are inadequate when we encounter speeds of immense magnitude. What are the terms and conditions in Newton's mechanics?

According to Newton, the metre-sticks and chronometers that I use to measure space and time are the same for you whether you are at rest or in uniform motion relative to me. Speed is the ratio of distance moved to time elapsed, and relative speeds add or subtract depending on whether you are heading towards or running away from a speeding object. That is all that was necessary to arrive at the conclusions about the two cars approaching each other at 30 metres per second. However, this 'common sense' fails for light beams, or for speeds comparable to that of light. Einstein realised that something must be wrong with our concepts of space and time. To crack the puzzle, he began to think about communication in a universe where light takes time to travel from source to receiver. In what sense, he asked, can we be sure that two events happen simultaneously? The answer is: we can't!

To understand why, imagine that you are in the middle carriage of a stationary train. You send a light signal to the driver at the front, and one to the guard at the rear. They will receive the signals at the same instant. Now, suppose that instead of being at rest, the train is moving at a constant speed. I am at the trackside, and as you pass me you send the light signals to the driver and the guard, as before. You will perceive them to arrive simultaneously – but I will not. The reason is that the light does not get there instantaneously. In the brief moment that the light beam travels from the middle to the ends of the train, the front of the train will have moved away from me while the rear will have moved closer. From my perspective, the guard receives the signal a few nanoseconds (billionths of a second) before the driver, whereas you will insist that they arrived simultaneously. By this 'thought experiment', Einstein realised that simultaneity as recorded by someone on the moving train is not simultaneity for someone at the trackside. In general, our definition of time intervals, the passage of time, depends on our relative motion.

If light travelled infinitely fast, as Newton supposed, there would be no problem with 'simultaneity'. In our daily experience the speed of light is effectively infinite, and these subtle properties of time are not noticed. Einstein had realised that the bizarre fact that the speed of light in empty space is a constant, independent of the motion of the transmitter or the receiver, somehow links with the notion of time differing for people who are in motion relative to one another. He worked out the logical consequences and found that not just time intervals but also distances, as measured in one inertial frame, differ from those measured in another, the mismatch depending on the relative speed of the two inertial frames.

Spatial intervals shrink, and time is stretched by a common amount. As you approach the speed of light, your perception of space and time relative to that of someone at rest is hugely different: distances are contracted to microscopic scales, while the metronome of time slows almost to a stop. Space and time themselves blend into an all-encompassing matrix referred to as space-time. At the speeds we are used to in our normal experience, space and time appear independent and these changes in space-time are so negligible that they are beyond our perception. Newton's mechanics was thus an effective theory of motion for velocities low enough to combine according to 'common sense'. Relativistic deviations were in the sixth place of decimals, at least for scientists up to the nineteenth century. For fast-moving atomic particles, however, such as those generated in the Large Hadron Collider at CERN, the effects of relativity are critical. In the final theory of everything, Einstein's relativity must be included.

4

Quantum theories of small things

Atoms in a hot body vibrate. They are electrical oscillators, emitting electromagnetic radiation whose mean frequency is proportional to the body's temperature. Thus you and I are visible to an infrared camera, which is sensitive to our body heat, but we do not radiate at higher frequencies: we do not glow in the dark. Hotter bodies do shine – first red, then yellow, and if the temperature is high enough, they become literally white hot – as they radiate throughout the spectrum.

That is what happens in practice, but it is not what Maxwell's theory predicted. According to Maxwell's theory of electromagnetism and light, the intensity in the frequencies well beyond violet, such as X-rays and gamma rays, should grow without limit, whatever the temperature. This singular failure of Maxwell's theory was a major shortcoming, not some subtlety buried in the sixth place of decimals. It was dubbed 'the ultraviolet catastrophe', and was one of the clouds that darkened Lord Kelvin's ideal sky in his 1900 speech.

That same year, the German theoretical physicist Max Planck found a solution. He postulated that the wobbling atoms in a hot body emit electromagnetic radiation as a staccato burst of discrete packets – soon to become known as quanta – like a rapidly dripping tap, rather than in a continuous stream as Maxwell had envisaged. By a combination of genius, persistence, and trial and error, Planck found a formula which describes how the intensity of radiation depends on both its frequency and the temperature. Its success depended on a critical, and ad hoc, assumption: that the energy of a single packet, or quantum, is proportional to the frequency of vibration of its electromagnetic fields. The constant of proportionality is now called Planck's constant and is denoted by the symbol h.

The theory of relativity, and now that of the quantum, temporarily cleared away Lord Kelvin's clouds, though new questions were already emerging. The quantum, for example, appeared to be a one-trick pony, designed to solve the ultraviolet catastrophe but having no further use. But in 1905 this changed when Einstein used the concept to show how it explains other phenomena, such as the ability of light to eject electrons from metals – the photoelectric effect.[1]

This was the second time that Einstein had seen mathematics as encapsulating deep truths about the physical universe. The mathematics of special relativity, for example, builds on equations known as the Lorentz–Fitzgerald contractions. These had been known for about fifteen years before Einstein saw that they were statements about the fabric of space-time, not mere mathematical tools. In Planck's mathematical formula, Einstein again saw a deeper truth. Planck had assumed that light is emitted in quanta called photons. Einstein now went further – light *is* photons. He assumed that the photons were real, in the sense that electromagnetic radiation is not merely emitted in staccato bursts, as Planck hypothesised, but travels like this. The radiant photons, for Einstein, are a fundamental property of light itself. Light consists of a stream of discrete particles: photons.

Quantum ideas were soon extended beyond light. If electromagnetic waves act like particles, so might material particles, such as electrons, act like waves. This proved timely: the confidence in the late nineteenth century that physics was 'complete' was being destroyed by a profusion of unexpected discoveries about the atom. In particular, atoms had been found to have an inner structure, comprising a compact massive central nucleus, positively charged, surrounded by lightweight, negatively charged electrons. According to Maxwell's theory, such atomic structures are impossible. If Maxwell's equations applied to electrons in atoms, then electrical attraction would cause the electrons to spiral down into the nucleus within mere fractions of a second. It would be utterly impossible for matter, as we know it, to exist.

The first step to the theory of atoms came in 1913 with the 'great Dane', Niels Bohr. While working in Manchester, Bohr took Planck's concept of the quantum and hypothesised that the angular momentum[2] of the electron in a hydrogen atom is an integer (whole-number) multiple of $h/2\pi$. This corresponds to saying that its angular momentum, when multiplied by the circumference of its orbit, is an integer multiple of h.

As those integer multiples increase, so does the energy of the electrons. The energy states thus increase stepwise, like ascending rungs on a ladder. An atom can emit electro-magnetic energy in the form of one or more photons if the electron drops from a high-energy rung to a lower one. This agrees with the observed discrete spectra emitted by atoms: the different colours of light in a spectrum correspond, in quantum theory, to photons of discrete energies.

The lowest energy level is when the integer is one. This corresponds to a radius of an electron orbit of about 1/20 of a nanometre, which is a measure of the size of an atom of hydrogen, which contains a single orbiting electron. An electron on this level cannot lose energy: there is no lower rung available for it to drop to. Thus in this state the atom can survive, in principle, for ever.

Like Planck's formula, Bohr's model was an ad hoc mathematical solution to a single problem. That these two would be the seeds of modern quantum theory began to become clear in 1923, when the French aristocrat and physicist Louis de Broglie developed a theory of matter waves. Where Planck had hypothesised that the frequency of electromagnetic oscillation is proportional to the energy of its particular quantum, de Broglie assumed that the wavelength of an 'electron-wave' is inversely proportional to the momentum of the associated material particle. This provided a physical basis for Bohr's atomic model.

We can visualise electrons as waves, like wobbles on a length of rope. When the rope is looped into a circle, like a lasso, the wave can fit perfectly into its circumference only if the number of wavelengths in a circuit is an integer. Electrons circulating in atoms can follow only those paths for which their waves fit exactly into the lasso. Thus Bohr's mathematical condition corresponds to a whole number of waves fitting along the circumference of the electron's orbit.

Quantum theory had been born with Max Planck in

December 1900,[3] and given physical reality with Einstein in 1905 and Bohr in 1913. Quantum *mechanics* – the more general theory of the dynamics of small things, with equations of motion that go beyond those of Newton – did not emerge until later. Louis de Broglie's hypothesis that electrons behave like waves inspired Erwin Schrödinger in 1925 to develop an equation of motion for such a wave. For the behaviour of the electron he introduced what he called a wavefunction, denoted by ψ. The probability of finding the electron at some location is proportional to the square of the magnitude of ψ. The magnitude of ψ oscillates in space and in time; Schrödinger constructed a differential equation which specifies its variation. Schrödinger's equation can be solved, at least in simple situations such as for the single electron circling a proton in a hydrogen atom, and describes the broad features of atomic physics well.

Just as Newton's laws of motion are the theory of macroscopic or 'classical' mechanics, so Schrödinger's equation is central to quantum mechanics. What they have in common is that they are differential equations, constructed to answer questions such as, 'If I specify the situation of a particle at a given time, what will be the situation at some other time?' Newton's classical mechanics determines the motion absolutely; Schrödinger's quantum mechanics only gives probabilities of this or that outcome. Thus classical and quantum mechanics have no obvious correspondence.

Like Newton's dynamics, Schrödinger's quantum mechanics applies only to situations where the rate of change of position – speed – is much less than that of light. In 1928 Paul Dirac discovered the quantum equation that describes the electron and conforms to the requirements of Einstein's special relativity theory. Dirac's equation treats space and time on an equal footing, as relativity requires. Also, when applied to an electron moving slowly compared with the speed of light, the solutions to Dirac's equation agree with those of Schrödinger.

The remarkable feature of Dirac's equation is that it has four interconnected parts, each of which is essential for the consistency of the whole edifice. That a single equation would be insufficient was already expected, because the electron empirically acts like a bipolar magnet. This leads to two possible quantum states when the electron is in an external magnetic field: the electron's magnetic moment can line up in the same direction as the field, or in the opposite direction. These two possibilities would require two equations, and Dirac's construction correctly describes the strength of the electron's magnetism, relative to its electric charge.

So far, so good. But what, then, is the significance of the extra pair of equations, leading to a quartet of linked parts? Dirac eventually deciphered them. To be consistent with relativity, it turns out that an electron cannot live alone. Instead, nature requires that there also exists a sibling of the electron: a particle with the same mass and magnitude of magnetism, but the opposite sign of electric charge, positive instead of negative – an example of what is known as anti-matter. This positron, or positive electron, was discovered in cosmic rays in 1932.

The four components of Dirac's equation thus describe the bipolar magnetic electron and its bipolar positron twin.

Here we see the uncanny ability of mathematics to antici-pate nature, to 'know' about reality before experiment reveals some phenomenon. Dirac was convinced that his equation was indeed the final piece in the quest for a theory of mechanics which incorporates both quantum ideas and relativity. He said:

> The underlying physical laws necessary for the mathematical theory of a large part of physics, and the whole of chemistry, are completely known. The difficulty is only that the exact applications of those laws leads to equations much too complicated to be soluble.[4]

Dirac may come across here as sounding as self-satisfied as Lord Kelvin, but note that his statement carefully delineates the boundaries. He makes claim for 'a large part' of physics, not all of it. It was fast becoming obvious that atomic nuclei have a labyrinthine inner structure, and Dirac said nothing about that. So far as chemistry and much of physics are con-cerned, in practice nuclear structure is irrelevant. Quantum mechanics itself also explains why this is so.

Atomic nuclei extend for at most a few femtometres (denoted by fm, equivalent to 10^{-15} m). Planck's constant, h, and the speed of light, c, provide a measure of the reach of the quantum universe and of special relativity.[5] Quantum uncertainty thus implies that energies of hundreds of millions of volts are needed in order to resolve structures such as nuclei, which exist on the scale of at most a few femtometres. Put another way, experiments carried out at room temperature, or even at the temperature of a Bunsen burner or a blast furnace, which only reach energies of at

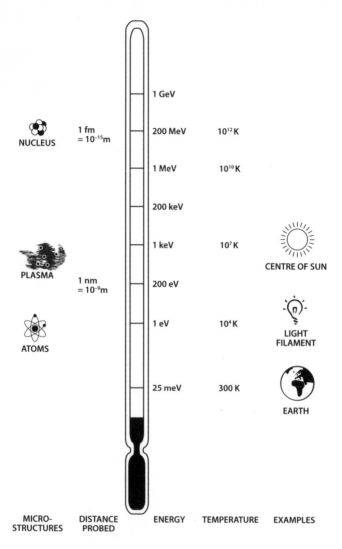

Figure 2: Below temperatures of millions of degrees, nuclear physics is in quarantine.

most a few electronvolts, are too feeble by factors of millions to be sensitive to nuclear structure (see Figure 2).

When New Zealander Ernest Rutherford discovered the atomic nucleus, in a series of experiments at Manchester conducted around 1910, he used beams of alpha particles as probes. The individual alpha particles, which are the nuclei of helium atoms, are the products of nuclear radioactivity with energies of just a few MeV. This is sufficient to reveal the presence of a lump of positive charge at the atom's heart, but nothing more. To discern the inner structure of the nucleus was beyond the energy reach at the time.

So, theories of phenomena that involve relatively low energies have no need for nuclear physics. In chemistry, for example, which deals with the interaction of atoms and molecules, nuclear physics is in quarantine. Interatomic forces are the result of electrical effects, such as the formation of electrically charged ions, which enable mutual electrical attraction to occur. This electrical activity all lies within the province of Dirac's equation. That chemistry survives as a distinct scientific discipline rather than being known as 'applied Dirac' is testimony to his final caveat: 'much too complicated to be soluble'.

QUANTUM ELECTRODYNAMICS

Maxwell had developed his theory of the electromagnetic field before the discovery of the electron, which carries electric charge, and of the photon, which transmits electromagnetic forces through the field. Quantum mechanics describes how the electron and photon behave, but it says nothing about how photons are created or absorbed by electrons. Thus Maxwell and quantum mechanics both fail to describe the most commonplace of phenomena: the creation and destruction of photons. Flick an electric light switch on, and photons pour forth in large numbers, perhaps to provide illumination for you as you read this. These photons will disappear when absorbed by electrons within the retina of your eye.

Dirac brought photons into the picture by creating a new type of theory: quantum field theory. His idea was that the entire universe is filled with an electromagnetic field. If energy is applied at some place and time, the excitation of this field results in the appearance of a photon. Conversely, the disappearance of a photon corresponds to de-excitation of the electromagnetic field. With this concept that electromagnetic fields can exist in discrete states of energy, Dirac had combined Maxwell's classical ideas with the quantum theory, and in a way that is consistent with the constraints of relativity. The resulting relativistic quantum theory of electric charge and light is called quantum electrodynamics.

Quantum electrodynamics, or QED as it is affectionately known, has become a paradigm for modern quantum field theories of the strong nuclear force, and of the weak force responsible for radioactive beta decay (more of which later). It has proved to be the theory of everything concerned with the electromagnetic field and how electrons interact with electromagnetic forces. QED showed that Dirac's equation, for all its beauty and apparent success, was but the first approximation to the theory of everything electronic. His equation had appeared to describe success-fully the spectrum of radiation from a hydrogen atom, and it also explained the size of an electron's magnetism – its so-called magnetic moment. Although these were remark-able advances on what had gone before, higher-precision data revealed deviations in the third place of decimals com-pared with what Dirac's equation implied. QED, however, explained these discrepancies.

Dirac's equation describes the electron interacting with a magnetic field, or the spectrum of photons radiated from within a hydrogen atom, as if these are happening in an otherwise empty void. In QED, however, the electron is also the source of an electromagnetic field, and that is pregnant with possibilities. In quantum theory, energy balance can be overridden for brief moments, enabling an electromagnetic field to spawn pairs of electrons and positrons which pop in and out of existence for infinitesimally short periods of time. These can in turn emit and absorb virtual photons, giving rise to an infinite number of possible ways in which energy can be shared among these various particles. The electromagnetic field is thus a vibrant medium, which QED describes correctly but which Dirac's equation ignored. That his equation was initially so successful, however, is thanks to a gift of nature.

The chance of one of these quantum fluctuations affect-ing a measurement, such as that of the electron's magnetic moment, happens to be small. For two or more fluctua-tions to affect a measurement, the chance is even smaller. Dirac's equation thus ignored quantities that are very small, but which become significant when measured with high precision. QED takes these contributions into account, as a series of terms in a formula whose magnitudes uniformly decrease. There is an infinite series of contributions. An infinite sum can have a finite value, however: for example,

$$1 + 1/2 + 1/4 + 1/8 + \ldots,$$

which sums to 2. For QED, a typical series is more like

$$1 + 1/100 + 1/10,000 + \ldots,$$

the sum of which is much smaller. The accuracy of your calculation is limited by your dedication to computing the precise values of ever-smaller terms in a series. This technique is known as 'perturbation': you calculate an answer (for example, by means of Dirac's equation) and then calculate the small correction that perturbs the original answer. It is by such techniques of perturbation calculations that the magnetic moment of the electron has been calculated in QED to an accuracy of more than nine decimal places.

For simple phenomena, such as a single electron interacting with photons of light, Dirac's QED is a theory of everything in that it enables calculations of quantities which can also be measured by experiment, and in some cases the agreement is as good as one part in 10^{12}. This is like measuring the width of the Atlantic Ocean to the precision of a human hair. The spectrum of light that emerges when an electron in a hydrogen atom jumps from one rung of the quantum ladder to another is also precisely as quantum electrodynamics implies it should be. This is also true for ions of other atomic elements, such as helium, whose nucleus (an alpha particle) has two units of positive charge: when one of the two electrons in a helium atom is removed, we have an ion. The behaviour of the remaining single electron is consistent with QED. But the problems of complexity arise almost as soon as we go beyond two electrons, to heavier elements, or try to compute the binding of a few atoms into even simple molecules. In practice, the equations are then insoluble.

These beautiful agreements between theory and data were only obtained after considerable time was spent trying to understand the mathematical significance of the infinite series. Initial attempts to add together the contributions of all the individual possibilities in the series of terms led to infinity as the answer. This is nonsense: 'infinite probability' is meaningless. Eventually, in 1947, a technique known as renormalisation was found to resolve the problem. In a nutshell, individual families of terms within the series were found to sum to infinity. Far from making the problem twice as hard, however, it turned out that one set which sums to infinity can be offset against another. The remaining terms give a finite result, which can then be determined by experiment.

If that were the whole story, though, QED would have no predictive power. It was discovered that, to eliminate the infinite sums, two empirical inputs are sufficient – they could be, for example, the value of the electric charge and the mass of the electron – and there are many other quantities that QED can calculate. Thus, for example, a first attempt at calculating the electron's magnetic moment gives an infinite answer, but once the sets of infinite terms related to the charge and mass are removed, and the constraint that a photon has no mass is imposed, the result is miraculously finite, and in agreement with measurements.

A quantum field theory for which only a finite number of such experimentally determined constants are required in order to obtain finite answers is said to be renormalisable. QED is renormalisable. Very few theories have this feature, and many theorists regard it as essential for constructing theories of everything. For a simple system – electron and photon – QED is indeed the theory of everything: and it works! Its limitations, as we have mentioned in general terms already, arise first from the difficulty in solving its equations for complex situations, and secondly from novel discoveries beyond the atomic level, such as in the nucleus. Most importantly, QED works for calculations in which there are no nuclear particles, such as finding the magnetic moment of the electron, or determining its interactions with light. If the electron in question is in an atom, the nucleus by definition is present. However, as long as nuclear physics remains in quarantine, with the nucleus treated as no more than a massive lump of electric charge, it does not infect QED.

However, a theory of truly everything must take us beyond this constraint. Dirac's theory explains the spectrum of photons emitted by electrons, but if the atom is hot enough, or if it is probed with sufficient resolution, for example by scattering high-energy electrons onto it, the nucleus is revealed to have a complex structure. It consists of protons and neutrons, which can move relative to one another and liberate energy in the form of gamma rays – photons up to a million times more energetic than those radiated from the peripheral electrons.

Today, physicists regard Dirac's construction of quantum electrodynamics as the theory of everything within atomic structure, but not of everything within the nuclear realm. Its empirical success has motivated the construction of theories of everything that extend the realm of application. The resulting theories – quantum chromodynamics (QCD) and quantum flavourdynamics (QFD) – incorporate features of nuclear physics and transmutation of particles and atomic elements, and are steps towards an ultimate theory of everything. They form the current standard model or core theory of particles and forces. This is a theory of everything, just as long as we quarantine quantum gravity.

THE HEART OF THE MATTER

Quantum theory implies that at temperatures in excess of a trillion degrees, or energies above 100 million electronvolts, distances of the order of a femtometre are resolved. Novel phenomena and structures of matter are revealed which go beyond anything that QED alone can describe. The structure of the atomic nucleus now comes into view. Its very existence immediately raises a paradox for the foundations of QED.

Nuclei are positively charged, dense clusters of protons and neutrons. A single positively charged proton is the nuclear heart of a hydrogen atom, but moving up the periodic table of the elements, all other atomic nuclei contain both protons and neutrons – collectively known as nucleons – in increasing numbers. Whereas the attraction of opposite charges entraps electrons in the outer reaches of such atoms, the repulsion of like charges should disrupt the nuclei of these elements, which contain several protons (92 in the case of the heaviest naturally occurring element, uranium). How can these protons remain together in a compact dense cluster?

The answer is that there must exist a strong attractive force acting among nucleons which is powerful enough to resist their mutual electrical repulsion and hold them in place. An explanation of this so-called strong nuclear force must be included in any theory that applies at such energies. It is absent from QED.

Atomic nuclei have a labyrinthine structure. Many are unstable, and attain stability by expelling some of their constituents, a phenomenon known as radioactivity. As a result, atoms of one element can transmute into those of another. Thus, for example, a uranium nucleus emits an alpha particle to become a thorium nucleus, and cascades down the periodic table by further radioactive decay until eventually stabilising as a nucleus of lead. The phenomenon of alpha decay fits naturally within quantum mechanics, and illustrates how nuclear physics still reads the quantum book: quantum mechanics and quantum field theory appear to be essential for theories of everything.

This is what happens. An alpha particle is a tight cluster of two protons and two neutrons, and is highly stable. Many large nuclei such as uranium, on the other hand, are barely able to resist the electrical disruption of their constituent protons. If some protons can be shed, along with their destabilising electrical effects, what is left can be more robust. Alpha decay is nature's most efficient way of doing this. However, first this quartet of nucleons has to escape the strong binding force of their colleagues in the uranium nucleus. It does so courtesy of quantum mechanics and the phenomenon of quantum tunnelling.

I once drew an analogy between quantum tunnelling and climbing over the Alps, from the Chamonix valley into Italy:

> The alpha particle initially is trapped within the valley. Classical physics would imply that the alpha particle will remain where it is – trapped inside the heavy nucleus – unless enough energy is supplied to climb up and over the mountain peak to reach the downward slopes of the far side. However, quantum mechanics allows it to escape by a process known as tunnelling. It is as if it can exit via the Mont Blanc tunnel, but only if it does so in a time less than that constrained by quantum mechanics.[6]

Once freed, the alpha cluster recoils violently from the highly positively charged residual nucleus, due to the electrical repulsion of like charges. Thus the phenomenon of alpha emission illustrates that both classical and quantum mechanics apply in the realm of nuclear physics.

There is another form of radioactivity, known as beta decay, in which a neutron converts into a proton (or vice versa). Electrical charge is conserved overall by the appearance of an electron (or positron) and a neutrino (an electrically neutral, almost massless sibling of the electron). This appearance of particles from the energy released in beta decay is a paradigm of quantum field theory at work.

But which is the relevant theory? Not quantum electro-dynamics, which does not describe processes in which electric charge moves around, or where particles change their identity. In QED, an electron remains an electron when it emits or absorbs a photon. In beta decay, by contrast, a neutron converts into a proton, and emits both energy and electric charge. The force involved in this case is known as the weak nuclear force, named 'weak' in contrast to the strong force, and also in recognition of its apparent feeble nature relative to the electromagnetic force.

The theory of everything that applies above energies in the MeV range will require a quantum field theory of the weak force. We also require a mathematical description of the strong force. These theories are in principle independent, but if there is truly a theory of everything into which they fit, we might expect them to have some features in common. And, if we are on the right track, the theories of the weak and strong forces should have some similarity with QED. This is indeed as it turns out.

QUANTUM FLAVOURDYNAMICS

In QED, the electrically charged electron interacts with a photon, to which it can transfer energy but not electric charge. The analogous quantum field theory of beta decay is almost the simplest possible generalisation of this: to allow the transfer of electric charge as well.[7] The resulting theory is called quantum flavourdynamics, or QFD. And it works.

The starting point is to treat the proton and neutron as

siblings in the sense that they are essentially identical but for their different 'flavours' – in this case electrical properties (and a difference in mass of 0.1 per cent). A similar twinning links the electron to the neutrino, which is in effect an electron with no charge, and almost no mass. In QED, photons interact with electrons; the neutrino, which is electrically neutral, has no role. In QFD, however, the neutrino is a leading actor.

There are also electrically charged siblings of the photon. These are conventionally known as W bosons, denoted by W^+ and W^-. The superscripts indicate their electric charges, which are identical in magnitude and sign to the electric charges of the proton and electron. An electron will turn into a neutrino when it absorbs a W^+; conversely, the emission of a W^+ by a neutrino turns it into an electron. The same is true for the proton and neutron: interaction with a W changes one into the other. In QFD, beta radioactivity arises when a neutron converts to a proton by emitting a W^-, which transforms into an electron and a neutrino (see

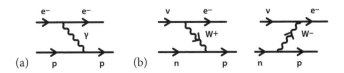

Figure 3: Particles and forces. The electron (e) and neutrino (ν), or proton (p) and neutron (n), feel the electromagnetic force (a) and the weak force (b) by exchanging respectively a photon (γ) or an electrically charged W boson (W^+ or W^-). The neutrino does not interact with photons, and does not feel the electromagnetic force.

Figure 4 on page 69). This transubstantiation of energy through the equivalence of energy and mass, expressed in Einstein's famous equation $E = mc^2$, is a key feature of quantum field theory.

This extension of QED into QFD successfully underwrites a wealth of phenomena, not least those involving neutrinos. Thus a simple generalisation of QED, which recognises the electron to be one of a pair of particles, reveals beta radioactivity to be a sibling of electromagnetism. They are not immediately twinned, however. The engine of beta radioactivity is dubbed the weak force in recognition of the fact that, as we saw earlier, it appears feeble relative to the electromagnetic force.

The 'weakness' is illusory, however. And though they seem very different at low energies, the electromagnetic and weak forces are like different aspects of a single 'electroweak' force. The difference in their strengths, as manifested in beta decay, is because the photon has no mass, whereas the W bosons are some 85 times the mass of a hydrogen atom. 'Weak' is thus a product of 'normal' electromagnetic strength which has been weighed down by the improbability that a single neutron can divest itself of 85 times its own energy. Quantum theory allows the energy account to go overdrawn, but the penalty for such profligacy is severe. This contrasts with the electromagnetic force, where massless photons are radiated without any such restriction.

The apparent difference in strength between QFD and QED is thus illusory, a consequence of the large mass of the W bosons. Calculations in QFD follow the same standard rules as in QED. And, as in QED, beyond the simplest approximation the quantum contributions add up to infinity. In QED, a finite physical answer emerged after renormalisation. This is possible because the photon has no mass, but in QFD the W boson is very massive. However, QFD is renormalisable, thanks to a further property of nature, linked to the Higgs boson.

The cosmos is filled with a so-called Higgs field, named after English physicist Peter Higgs, who developed the hypothesis in 1964. That this is real was confirmed in 2012 when experiments showed that 125 GeV of energy can excite the particle quantum of the Higgs field – the Higgs boson. The relevance of the Higgs field is that the masses of fundamental particles such as the electron and, crucially, the W bosons arise from interactions with this universal background field. With the Higgs field included in the cast, QFD is renormalisable, and its finite predictions agree with experiment. Thus, immersed in this Higgs field, we have a theory of everything to do with the so-called weak force. (More about the Higgs boson at the end of this chapter.)

There is also a huge bonus in the quest for an ultimate theory of everything, in that the mathematical patterns of QFD are remarkably similar to those of QED. This hints that electromagnetism, light and also radioactivity are themselves profoundly linked in theory. This tantalising glimpse that physicists were now on a journey towards a universal theory of everything was reinforced with the discovery of the viable, renormalisable, quantum field theory of the strong nuclear force: quantum chromodynamics, or QCD.

QUANTUM CHROMODYNAMICS

The phenomenon of radioactivity is literally radiated into our surroundings. Radium is warm to the touch; localised heat in a cool background can indicate the presence of radioactivity. The atomic nucleus, however, is invisible at room temperature. Even when an atom is irradiated with the warm energetic beams that radioactivity has gifted us, the atomic nucleus appears as nothing more than an inert passive lump of electric charge. In such circumstances, nuclear dynamics can be consigned to quarantine. QED and QFD rule.

At temperatures of billions of degrees, however, where particles have energies in the MeV range, nuclear structure becomes apparent. It consists of neutrons and protons bound to one another by the strong force. The key to the theory of this force is revealed when the energy scale is several thousand times bigger – from tens to hundreds of billions of electronvolts (in the GeV range). This scale of energy reveals phenomena at distances much smaller even than the breadth of a proton or neutron – and these nuclear constituents are themselves found to have structure. They are clusters of trios of even more fundamental particles: quarks.

The quark level of reality is the lowest currently accessible to experiments. Whatever structures exist at distances smaller than about one ten-thousandth the breadth of a proton, or at energies in excess of ten trillion electronvolts (10 TeV), they remain out of our reach, at least for now. In any event, it appears that any such lower level of structure can be quarantined for our immediate purposes: the quark layer of the cosmic onion is key to the quantum field theory of the strong nuclear force.

Quarks in protons or neutrons come in two so-called flavours: up and down. They carry electric charges, which are fractions of that of a proton or electron. The up quark has charge $+2/3$ and the down has $-1/3$, so two ups and one down comprise a proton, while two downs and one up give the neutral neutron. Combinations of three ups or three downs form short-lived particles, which have been seen in experiments in particle accelerators such as at CERN, but they need not concern us here.

The beta decay of the neutron is revealed to arise from a more fundamental reaction at quark level: a down quark converts to an up quark, the spare energy materialising as an electron and an antineutrino (the antimatter version of a neutrino). Its partners, an up and a down, act as mere

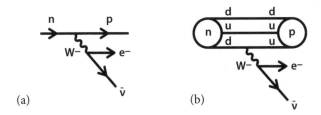

Figure 4: Beta decay. The beta decay of the neutron (seen in a) is due to a fundamental transmutation at the quark level. This involves a down quark (labelled d) converting to an up quark (u), in the process emitting a negatively charged W boson (W⁻) which then converts to an electron and an antineutrino (v̄) (seen in b).

spectators, the change in that down quark being enough to convert the neutron into a proton. The beta decay observed at the quark level turns out to be identical to when an electron converts to a neutrino, or vice versa. As far as QFD is concerned, the electron–neutrino duo are the same as an up and a down quark (see Figure 4). Their electric properties are distinct, of course, as the magnitudes of their electric charges differ: 0 for the neutrino, 2/3 for the up quark, −1/3 for the down quark and −1 for the electron. But here too there is a profound symmetry: the response of each of the quarks and the electron to both electric and magnetic fields is in direct proportion to the magnitudes of these charges.

We see here hints of some symmetry between the seeds of nuclear matter – the quarks – and the electron and neutrino. A further remarkable property is that electric charges balance in matter overall. A hydrogen atom, for example, has the charge of its negative electron precisely balanced by the positive proton, but in a complex way: quarks cluster in threes, not twos or fours, for example, yet their one-third fractions of electric charge miraculously add up so as to cancel out. Later, I shall examine this clue to the structure of an ultimate theory of everything. But first, there is another clue: the strong force reveals why quarks form such convenient triplets.

In addition to their electric charges, quarks carry another attribute, known as colour charge, or simply colour. Just as electric charge is the source of the electromagnetic force, so is colour the source of the strong force. This novel quark charge comes in three varieties, usually labelled red, blue and green. And just as Dirac required the electron to have an antiparticle – the oppositely charged positron – so his equation implies that an antiquark mirrors each quark. Thus the anti-up and anti-down quarks have the opposite sign of electric charge to their material counterparts, and also the opposite sign of colour, thus red, blue or green, but this time negative.

The rules of combination for colour charges match those of electric charges: like colours repel, while unlike or opposite attract.[8] Thus three quarks, each of a different colour, mutually attract to form a trio, as in the proton or neutron. A quark and an antiquark with positive and negative colours will also mutually attract. These short-lived states, known as mesons, are produced transiently in collisions between cosmic rays and the atmosphere, or in experiments in particle accelerators. Thus the rich behaviours we see on the femtometre scale, such as the existence and dynamics of atomic nuclei, are caused by trios of coloured quarks at a deeper level.

Such phenomena are manifested only at energies of many billions of electronvolts (GeV), far beyond the conditions we normally experience. In moving from atoms to quarks, we have progressed through some ten orders of magnitude, in size or energy, and the new physics appears to be controlled by a theory which is like an extension of QED. The main difference is one of strength but, as we shall see later, even this is understood.

The similarity between colour and electric charge is real, and profound. Just as combining electric charge relativity and quantum field theory leads to quantum electrodynamics, so does the introduction of colour charge lead to a similar mathematical description. The resulting theory is known as QCD, for quantum chromodynamics. The parallel is compelling. The quantum particles of the electromagnetic field in QED are massless photons; in QCD we are inexorably led to massless gluons. The name of these particles reflects their function: they 'glue' quarks tightly together in trios which we recognise on the femtometre scale as protons or neutrons.

For half a century after the discovery of the atomic nucleus, its structure was envisioned simply in terms of neutrons and protons. These nucleons were assumed to play the main roles in theories of nuclear structure and nuclear reactions. Empirical rules were determined to describe phenomena at energies from millions to billions of electronvolts (MeV to GeV). From about 1970, however, when the frontiers of high-energy physics reached energies of hundreds of GeV, the inner world of the nucleons was revealed and the quark layer of reality came to centre stage. In the decades since then, experiments have shown that QCD is the correct theory of the strong interaction.

Like its older cousins, QCD is renormalisable. At very high energies, where the role of QCD is most clearly revealed, the affinity of quarks for gluons is found to be similar in strength and identical in behaviour to that of electrons for photons in QED. The mathematical techniques that worked so well in QED, such as making successive approximations (perturbation theory) also apply in QCD, though with some differences in technical detail.

One of the implications of QCD is that the strength of the force depends sensitively on the energies and distances involved. At distances of roughly one femtometre, the affinity between quarks and gluons is so great that these constituents are unable to escape from their prison. Permanently confined within nucleons, quarks become invisible on the MeV scale of nuclear physics. This strong affinity is the source of the powerful attractive force that binds atomic nuclei. Thus although the equations of QCD are complex, and have only been solved for a limited number of cases, the qualitative behaviour of nuclear physics is seen to emerge from this fundamental underlying dynamics.

At ever shorter distances and extremely high energies, QCD predicts that quarks and gluons couple less powerfully. Such behaviour has been confirmed in experiments at the Large Hadron Collider. If this prediction proves also to be true at even higher energies than those accessible at the LHC, it will have an exciting implication: if we could examine matter at distances as small as 10^{-30} m – in a trillion times finer detail than we can at present – the strengths of the electromagnetic and colour (strong) forces would become similar. This is a compelling sign that in QED and QCD we, from our still relatively cool environment, have glimpsed an outline of a unified theory of the electromagnetic and strong forces. Their similarity with QFD is a further sign that here is a key to the ultimate theory of everything – or at least, of 'everything-without-gravity'.

THE HIGGS BOSON: CAPSTONE OF THE CORE THEORY

The predicted presence of a massive W boson initially clouded the otherwise perfect trinity of viable quantum-field theories for the atomic and nuclear forces. This cloud evaporated in 2012 following discovery of the Higgs boson, which is now the capstone of the current core theory. The origins of the breakthrough go back half a century, however, to an intriguing phenomenon in QED. The quantum mathematics of QED imply that if no other fields fill the vacuum, then the mass of the photon is zero – in accord with common experience.[9] There is a circumstance, however, where the photon is observed to act as if it does have mass: when it is in the presence of plasma. This stimulated a thought experiment, in the best traditions of Albert Einstein, which will lead us to the idea of the Higgs boson.

More than a hundred kilometres above our heads is the ionosphere, where radiation from the Sun and outer space splits atoms into negatively charged electrons and positive ions. This state of matter is known as plasma. The best-known property of the ionosphere is the way it affects the propagation of radio waves.

Low-frequency waves cannot exist in a plasma (see Figure 5). When AM radio signals reach the lower edge of the ionosphere, they may be reflected, like light from a mirror. Having been turned back towards the ground, they hit the surface of the Earth and bounce upwards, only to be reflected back by the ionosphere again. (This zigzag propagation is what allowed the first long-distance radio

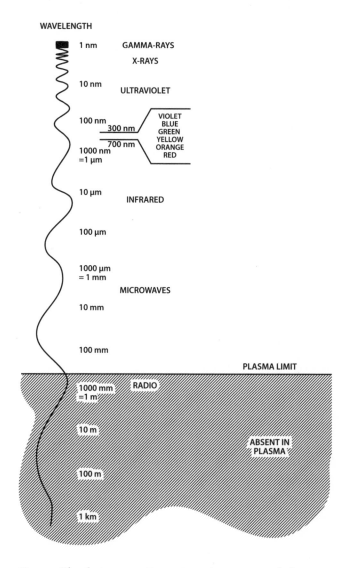

Figure 5: The electromagnetic spectrum in vacuum and plasma.

communication.) Visible light consists of electromagnetic radiation of higher frequency than that of a radio wave. Although the ionosphere rejects the low-frequency radio waves, it is transparent to visible light – we can see the stars even if AM radio waves cannot pass through it. So the key feature of a plasma is that electromagnetic waves can penetrate it only if their frequency is higher than some critical amount, known as the plasma frequency or the plasma limit.

Thus it is possible for the frequency of a radio wave to be less than the plasma frequency, while that of visible light exceeds it. The property that only waves of high frequency can pass through a plasma, while those below the plasma frequency are cut off, gave the American theorist Philip Anderson in 1962 his insight about the subtle ways in which photons of light can sometimes act like massive particles. He imagined how a creature that lived inside a plasma would experience electromagnetic radiation and photons. The creature would only ever be aware of electromagnetic radiation oscillating faster than the plasma frequency. Planck showed that the energy of each photon in an electromagnetic wave is proportional to the frequency of the wave. Thus a minimum frequency corresponds to a minimum energy: photons in plasma have a minimum energy, which is not zero.

Now, a massless photon can have any energy, all the way down to zero. Contrast this with a massive object: it can reduce its kinetic energy to zero, by coming to rest, but an intrinsic amount of energy remains (given by mc^2). Thus the minimum energy of a particle is a measure of its mass. In empty space a photon can have zero energy, consistent with its zero mass. In a plasma, however, the photon acts as if it does have mass. From the perspective of Anderson's hypothetical creature, which is unaware that it is embedded in a plasma, the photon *is* massive. We can imagine that some Paul Dirac in this mythical environment would develop quantum electrodynamics – with a massive photon.

Here comes the conceptual leap. We are like this creature: we are immersed in some all-pervading field, which we may call 'electroweak plasma'. What it consists of is, for now, unknown. Its property, however, is that it is transparent to photons (which are thus massless) but not to W bosons (and other particles). The W bosons appear massive to us, living as we do in this weird electroweak plasma and ignorant of its presence. In the absence of this electroweak plasma, the particles would have remained massless. The implication is that we perceive the W boson and other fundamental particles to have mass because of their interaction with this ubiquitous field.

This was a nice theory, but it was fifty years before an experiment was able to confirm it. If just the right amount of energy is applied, a plasma will resonate. In quantum theory, the plasma resonance acts like a particle, known as a plasmon. A similar idea can be applied to the electroweak plasma: just the right amount of energy will excite the electroweak plasmon – in other words, the Higgs boson. If this could be demonstrated experimentally, the presence of the electroweak plasma would be confirmed.

This may sound as if we have re-introduced the ether, a fluid once supposed to be responsible for the propagation of electromagnetic waves, but was famously eliminated by Einstein with his theory of relativity. What Peter Higgs and others did in 1964 was to develop this conceptual idea in a form that is consistent with special relativity. And in 2012, the Higgs boson was discovered, confirming that we are indeed immersed in an electroweak plasma – the Higgs field. We currently have no knowledge of what this stuff might consist of. We just know that it exists and that it takes about 125 GeV of energy to reach the small distances of some 10^{-18} metres at which particles interact with it.

Hopefully, as we learn more about this electroweak plasma, we will get insights into why interactions with it lead to the observed range of particle masses and the relative strengths of the fundamental forces. Although our knowledge is still far from complete, it is nonetheless plausible that we have here the theory of everything needed to describe the nature of matter, at least that of which we are made, at the energies accessible to modern experiments.

As for a theory of truly everything, that is still remote. And in everything that we have looked at so far, we have not confronted the marriage of quantum field theory and general relativity – the subject of the next chapter.

5

Weighty matters

A UNIVERSE OF ENERGY

The fundamental building blocks of a modern theory of everything include quantum mechanics and a founding principle of Einstein's special relativity theory: that the speed of light is a universal constant. The American theoretical physicist Steven Weinberg has argued that a final theory will contain these two features. I agree with him, if only for lack of any better insight, yet until the start of the twentieth century science had progressed without apparent need for either of them.[1] It goes to show that successful theories encompassing vast swathes of phenomena can be created, even while fundamental pieces of the theory of truly everything are absent. The historical examples we have looked at so far give clues to why this happens. Weinberg's insight clarifies not only the robustness of the modern core theory, but also the deficiencies that a final theory must address.

Let us go back to Newton's laws of motion. From the seventeenth century, they described the motion of the planets, the Moon, and everyday objects on Earth, in agreement with all experimental data, for some two hundred years. Had modern experimental apparatus been available to Newton and his peers, they would have detected deviations at the sixth place of decimals which today we would understand to be due to the effects of relativity. The reason why Newtonian physics was successful for so long, and continues to be of immense practical use, is because the speed of light, some 300,000 km/s, is huge compared with the speeds of objects studied by all scientists and engineers before the twentieth century, and in practice for most situations even today.

Here we see a key rule: an effective theory can work in practice when some crucial parameter in the 'final theory' is on an utterly different scale to the magnitude of the actual quantities being measured. Thus the speed of an object relative to that of light is the key to Newton's success, notwithstanding that nature is more completely described by Einstein's richer theory.

When looking for such a key parameter, we must compare like with like, of course. An elephant is not large compared with a second, nor is a kilometre small compared with the speed of light. The key is that we have to take account of dimensions. So an elephant is large compared with a flea, and a kilometre likewise relative to the diameter of an atom of hydrogen. In the case of the speed of light, we can meaningfully compare its scale against quantities with the dimension of length per unit time: some number of kilometres per second relative to the 300,000 km/s that we denote by c.

Next, let us explore the other boundary of Newton's landscape: the need for quantum mechanics. Newton's mechanics works empirically when applied to large objects (which also, of course, must move slowly). Quantum mechanics is essential to any description of atoms and small things. But 'small' relative to what, exactly? In the case of speeds, where Newton's laws work because the riches of special relativity theory may be ignored, the crucial parameter, speed, has dimensions of length per unit time. For quantum mechanics, the crucial parameter is Planck's constant, denoted by h, which is measured in units of energy multiplied by time: joule-seconds, say.

Such a combination – known as the action – is rarely met with in everyday life, but it comes into the picture whenever we analyse the dynamics of atoms and their constituent particles. So, it is the size of an object's action relative to the magnitude of h that determines whether quantum mechanics needs to be considered. When the magnitude of the action is smaller than or similar to h, quantum theory applies and Newton is inadequate. When it is large relative to h, however, corrections to Newton's theory will again be in the sixth place of decimals.

A reason why science was unaware of these limitations to the received wisdom of the nineteenth century becomes clear when we combine h and c to form the product hc. The dimension of h is energy multiplied by time, and c is length divided by time, so their product has dimensions of energy multiplied by length. This combination might also appear strange, but it is fundamental to physics and provides a measure of when the classical theory of Newton is an effective theory of everything.

The magnitude of hc in joule-metres is some 2×10^{-25} – that is, two tenths of a trillionth of a trillionth, which is unimaginably small. The units of joules for energy and metres for distance, which are so well suited to pre-twentieth-century classical science and engineering, are clearly on a scale completely inappropriate to the magnitude of *hc* and the quantum world of atomic physics.[2] The significance of *hc* for our ability to understand nature is also that it sets the scale of how much energy you must expend in order to resolve matter at small distances.

Although many of us pay our energy bills according to the number of kilowatt-hours we have used,[3] we get a more immediate sense of energy from the heat or cold associated with temperature. As we have seen, one of the great insights of nineteenth-century science was that the energy of particles in motion is perceived as heat. The resulting branch of physics, thermodynamics, is well named: it deals with the dynamics and motion of heat. If a vast number of particles bump into one another, they will transfer energy from one to the next until the whole collection comes to equilibrium. Some will be moving faster than the average, others slower, so we cannot say that a given temperature corresponds precisely to a specific energy. Nonetheless, for our purposes it is sufficient to say that room temperature, about 300 K, corresponds in energy terms to about 4×10^{-21} joules.

Temperatures of several thousand degrees, at which objects begin to glow visibly, correspond to energies that are about ten times higher than room temperature. When a hot body at a temperature of thousands of degrees radiates light, the wavelength of that light is of the order of a micrometre or less. In fact, visible light ranges from about 400 to 750 nanometres. That is the practical limit for resolving images by means of the visible spectrum. A bigger microscope

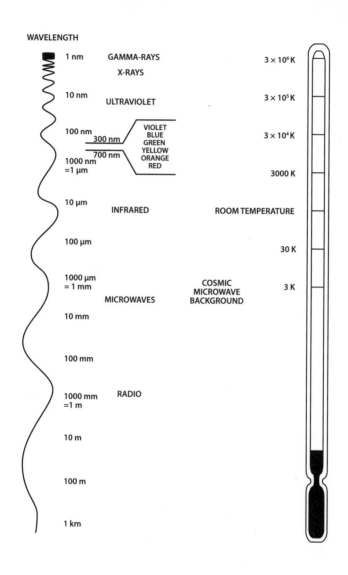

Figure 6: Temperature, distance and the electromagnetic spectrum.

might enlarge the image, but the wavelength of the radiation – light – limits the resolution (see Figure 6). The energies accessible by scientists in the nineteenth century were very limited compared with what modern technologies provide. In consequence, those scientists were limited in their ability to resolve the micro-world.

Nowadays we have access to powerful machines that can focus beams of particles whose individual energies are some ten trillion times larger than those of photons of visible light. The Large Hadron Collider at CERN is an example. A power plant liberates energy, which can be delivered to instruments in the LHC. These create electric and magnetic fields, which thrust protons into motion. Accelerated to almost the speed of light, beams of these protons are made to collide head on. The resulting collisions of individual particles are at energies up to a hundred trillion times larger than the energies at room temperature. As a result, these collisions can reveal structures in nature which are correspondingly smaller than our eyes can see – up to a thousand times smaller even than the nucleus of a hydrogen atom.

Under such conditions, quantum mechanics and special relativity are absolutely necessary to describe the phenomena. However, it is because such conditions are so far removed from the relatively low energies available to scientists of the nineteenth century that classical theories, such as those pioneered by Newton, seemed to be candidates for a theory of everything. Albert Michelson was not far from the truth when he mused that discoveries could be found at the sixth place of decimals; all that was lacking in practice was experimental apparatus sensitive enough to make such fine measurements. For all practical purposes, experience was limited to phenomena restricted to a relatively low region of energy, and Victorian science, unaware of the labyrinthine inner workings of the atom, possessed an excellent working theory of everything then known.

GRAVITY IN QUARANTINE

For centuries, science was able to create very successful theories while remaining unaware of the need for quantum mechanics or relativity. At the start of the twenty-first century we are in an analogous situation as regards gravity. Our core theory contains quantum chromodynamics, which underpins the existence of atomic nuclei, quantum flavourdynamics, which describes phenomena such as the transmutation of elements via beta decay, and quantum electrodynamics. This trio are so similar mathematically that they are probably subspecies of some theory of every-thing which describes the dynamics of atoms, atomic nuclei, protons, neutrons and their constituent quarks at energies beyond those accessible by the LHC. None of this says any-thing about gravity, however.

Massive bodies attract one another with a strength that dies away as the inverse square of the distance between them, a property that has been known for centuries. At cosmic scales, gravity is the most noticeable force – and yet our core theory ignores it. Why is it possible to quarantine gravity?

Gravity is exceptional in that it acts attractively between all particles. The force of gravity is thus cumulative in that each individual atom adds its own trifling gravitational attraction to that of all other atoms. The Sun is huge, which is why its gravitational pull is felt throughout the solar system. The electromagnetic force can attract or repel, and in bulk matter these competing effects tend to cancel out, leaving gravity prominent on the cosmic scale. The strength of gravity at the level of individual atoms, however, is negligible compared with, say, the strength of the electric and magnetic forces: a small magnet will stick to the door of your fridge, overcoming the gravitational pull of the entire planet.

For a quantitative measure of gravity's intrinsic feebleness, consider an atom of hydrogen. This consists of a single electron and a proton. These particles carry negative and positive electric charges respectively, and mutually attract by an electric force which fades away as the inverse square of their separation. The same holds true for the force of gravity. The electron and proton each have a mass, so they mutually attract gravitationally. This attraction also dies off as the inverse square of distance, so in this case we can make a direct comparison of the relative contributions of the electric and gravitational forces. What we find is that the force of gravity is some forty orders of magnitude – that's 10^{40} – smaller than its electric counterpart. So gravity is utterly negligible on the atomic scale, and our core theory works. To identify any deviations from the core theory's predictions would require measurements whose sensitivity is far beyond even the sixth place of decimals.

This modern experience has parallels with the situation at the end of the nineteenth century. Today we ignore gravity, but we nonetheless have a successful core theory. Back then, the classical theories of mechanics and thermo- dynamics worked well, even though the need for relativity and quantum mechanics lurked in the wings. The quanti- tative reason, as we saw, was exposed by the magnitude of the product $\hbar c$ (where $\hbar$ is h divided by 2π), which requires energies more typically found in stars than on the Earth in order to release quantum mechanics from quarantine. What is the corresponding measure for gravity?

Newton's law of universal gravitation says that the force between two massive bodies is proportional to the product of their masses divided by the square of their separation. This expresses how the force differs from one body to another, or varies across space, but it says nothing about its absolute magnitude. That requires a further scale, which is known as the gravitational constant, denoted by G. It is the smallness of G that controls the intrinsically feeble strength of gravity. Its value is some 6.6×10^{-11}, with units of cubic metres per kilogram per second per second.

This dependence on mass, length and time may seem bizarre, but it turns out that when we compare it with the dimensions of $\hbar c$, something rather simple emerges. If we divide $\hbar c$ by G and take the square root, we get a quantity with the dimensions of mass. This mass, which has a value of some 10^{19} – that's ten million trillion – times that of a hydrogen atom, is known as the Planck mass, after Max Planck, who first introduced the concept of the quantum into physics, and in so doing first noticed this combination of h, c and G (see Figure 7).

The magnitude of the Planck mass is similar to the mass

of a single human hair. A hair contains trillions of atoms and so, relative to the mass of a single atom, the Planck mass is huge. The gap between the energies of room temperature and those accessible at the Large Hadron Collider is a similar power of ten to the gap from the LHC to the Planck scale. If we were suddenly shown phenomena that occur at energies on the Planck scale, in all probability they would be as unfamiliar to us as quarks, gluons and the Higgs boson would have been to Newton.

The Planck mass is large because G is small. The remoteness of the Planck scale of energy compared with the most extreme energies of experiments within our reach today is the physical reason why our core theory, which ignores gravity, works so well when applied to atoms and their constituent particles.

We might imagine some super-LHC which in the future could perform experiments at energies close to the Planck scale. Under such conditions, as we shall see later, the effects of gravity on individual particles could no longer be put into quarantine and ignored. If quantum mechanics is

PLANCK ENERGY/MASS

$$\sqrt{\frac{\hbar c}{G}} \simeq 1.25 \times 10^{19} \text{ GeV}$$

PLANCK LENGTH

$$\sqrt{G\hbar c} \begin{aligned} &= 1.6 \times 10^{-35} \text{ m} \\ &= 1.6 \times 10^{-20} \text{ fm} \end{aligned}$$

PLANCK TIME

$$\sqrt{\frac{G\hbar}{c}} = 0.5 \times 10^{-43} \text{ sec.}$$

Figure 7: Planck scales: mass, length and time.

fundamental to the final theory of everything, then a viable description of particles subjected to the force of gravity, consistent with quantum mechanics, will be paramount. A final theory of everything must take such phenomena into account.

The remoteness of the Planck scale is thus both a blessing and a curse. On the one hand, our core theory has no need to include quantum gravity because any deviations that result from its neglect are utterly negligible. The downside, however, is that the domain of quantum gravity is so far away that it is impossible for us to find any clues to guide us towards the correct theory.

GRAVITY BEYOND NEWTON

In 1905, Einstein showed that in a universe without gravity, Newton's classical mechanics is an approximation to a richer theory, which we know as special relativity. Newton's theory is an excellent approximation so long as we limit our attention to objects whose speed is small relative to that of light.

Einstein's theory revealed also that a massive body at rest contains an amount of energy given by $E = mc^2$, where c is the speed of light. When the body is in motion, its total energy is the sum of this rest energy, as it is known, and the energy of its motion, its kinetic energy. So long as the total energy and the rest energy are nearly equal, Newton's classical theory is an excellent approximation to Einstein's special relativity theory.

This is the case when the body's speed is small compared

with that of light. In high-energy physics, as its name suggests, this is no longer true. Protons in the LHC, for example, are accelerated so near to light speed that their energy is more than a thousand times larger than their mc^2, and Newton's dynamics is completely inadequate for describing their behaviour. For particles at high energy, Einstein's special relativity theory holds sway.

The label 'special' indicates that this theory of Einstein's was specialised to the case where gravity is negligible. For the behaviour of individual protons and electrons this is true in practice, and as a result the core theory of particle physics employs special relativity with great success. But what if Einstein had not made this special restriction? How does Newton's concept of a gravitational field marry with the profound visions of nature that special relativity revealed?

In special relativity, Einstein had not simply demonstrated that Newton's equations are approximations to some richer mathematics. He had changed our perception of space and time. In Newton's dynamics, we move through some invisible three-dimensional matrix of space and measure the passage of time by the steady metronome beat of an imaginary clock. This four-dimensional structure (the three familiar dimensions of space, plus time) remains fixed, unchanged, as we pass through it. Einstein's special relativity, however, implies that our motion subtly alters its fabric.

Someone who moves at a constant high speed relative to you will perceive the spatial grid to contract, and the metronome beat to slow relative to how you, at rest, experience the passage of time. Steady motion, in the sense that we do not speed up or slow down (in essence, that no external forces push or hinder us), alters the fabric of space-time. Here is the essential key to special relativity theory: the special theory is a statement of how space-time adapts to uniform motion in the absence of forces, such as gravity. Einstein thus had an explicit problem to solve: how does gravity fit into this new view of space-time?

The axiom that the velocity of light is a universal constant led Einstein to special relativity. Yet gravity acts on energy in all its forms, and as photons have energy, gravity will deflect a light beam. Furthermore, gravity fills the cosmos, so light beams are universally disturbed. Einstein therefore had a conundrum: how could a theory of gravity satisfy the principles of relativity, which assumed that light travels in straight lines at constant velocity? It seemed that the fundamental postulate of relativity could survive only if the effects of gravity could somehow be turned off.

Einstein's revelation was that gravity *is* effectively switched off for a body in free fall. What you perceive as the weight of a heavy object is the force you have to apply to stop it falling to the ground. As for your own heaviness, a weighing machine measures how much force it has to exert to prevent you falling to the centre of the Earth. If the floor and the Earth were just vapour, we would all fall to the centre of the planet, weightless.

Einstein had come to his original theory of special relativity with the assumption that there is no absolute state of rest. His general theory arose from the idea that there is no absolute measure of force and acceleration. A century later, we can see this in videos of astronauts floating weightless in the cabin of the International Space Station. The ISS and its human cargo are all in free fall, the 'floor' beneath the craft being nothing but thin air. They are also travelling 'horizontally' at such a speed that the Earth's curvature makes the ground fall away at the same rate as they are descending towards it.

The astronauts, meanwhile, have no sense of any gravitational force. Objects in the cabin, which are all falling at the same rate, appear to remain in a state of suspended animation, as if they feel no force. Within the sealed compartment of the ISS, astronauts have every right to regard themselves as being at rest. Newton's law of inertia implies that, in effect, gravity has vanished in a free-falling, weightless state.

This also applies to beams of light. Suppose an astronaut shines a torch horizontally relative to the floor of the ISS. In the cabin, the beam of light travels in a straight line. In the few nanoseconds of its flight, however, the ISS and the photons of light have fallen towards the Earth. A precision measurement by someone on Earth would detect that the light beam had bent subtly as it 'falls' under gravity.

This is a twenty-first-century analogue of what Einstein had visualised when he thought about general relativity. Next he had to build a mathematical description. His inspiration for this came when he considered a collection of objects in free fall. We can imagine a convoy of spacecraft, at rest relative to one other, but all falling under the influence of the Earth's gravity. After a while, the astronauts in them would begin to notice that all the spacecraft were getting nearer to one another. This is because in free fall their trajectories are all converging towards a point at the centre of the Earth.

Einstein's inspiration was his analogy between this picture of freely falling objects and the convergence of lines of longitude at the north and south poles of the Earth. Mercator's projection of the Earth's surface onto a flat map represents all these lines as parallel. This is a fair approximation near the equator, but as one heads northwards they gradually converge, eventually all coming together at the north pole (see Figure 8). The reason is that the two-dimensional surface of the Earth is curved in a third dimension.

Einstein used this to make an analogy with gravity: the curving trajectories in free fall are like lines of longitude on a 'surface' that curves in some higher dimension. A popular interpretation of this analogy is that the three-dimensional 'surface' of space is stretched by large masses. In reality, the requirements of relativity led Einstein to a mathematical theory of gravity – general relativity – where space-time is warped by the presence of momentum and energy, not just of mass.

The shortest distance between two points on a flat surface is a straight line. If you move at constant speed, with no forces acting on you, this is the path that takes the least time. The same is true in general relativity: the path between two points followed by material bodies, and also by light beams, is the one which takes the shortest time.

This is familiar in optics when a ray of light travels through different media. The path of least time leads to the bending of light when it crosses the boundary from one

(a)

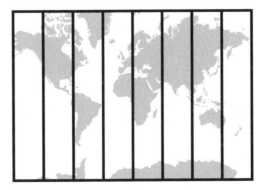

(b)

Figure 8: Lines of longitude and curved space. (a) On a curved surface, all lines of longitude converge at the poles. The trajectories of bodies converge at the centre of gravity (centre of mass) of the attractor. (b) On a flat (Mercator) projection, the equatorial regions are accurate but the polar regions are distorted. This is analogous to flat space-time being distorted in the presence of a strong gravitational force.

medium to another, which we call refraction, as when a stick dipped into water at an angle will appear to be bent. In the rainbow of light, different colours correspond to different rates of oscillation of the electromagnetic field, each of which seeks its shortest optical path when it meets an interface between air and glass, as in a prism. And so it is in space-time for bodies: a comet rounding the Sun is following the path that minimises the time it takes to pass from deep space on one side of the solar system to far away on the other.

If Einstein was on the comet, he would insist it was in free fall, effectively at rest and free of forces. Newton, by contrast, would have interpreted the comet's trajectory as a curved orbit governed by the gravitational force of the Sun. The link between these viewpoints is, of course, that the gravitational field of the Sun is what has warped space-time. To build his theory of gravity, Einstein thus had to answer two questions. First, how does some particular distribution of matter distort space-time? And, having determined the form of this warping, how do bodies then move around? In addition, he had to ensure that the constraints of special relativity are satisfied.

The first step involves Einstein's equivalence between energy and mass, $E = mc^2$. Newton's theory implies that the gravitational interaction between two bodies which are initially at rest relative to one another is proportional to the universal scale of strength, G, and to the product of their masses. Einstein's $E = mc^2$ thus generalises this interaction to be proportional to G, and the product of the individual energies, all divided by the fourth power of the speed of light, c^4.

We have considered the two bodies to be at rest, but their mutual interaction will cause them to move towards one another. This will give them both energy and momentum. Einstein's experience with electromagnetism and the foundations of special relativity provided the key to this complexity. We have seen that electric and magnetic fields, or space and time, are only cleanly separated with respect to one observer; for another observer in motion relative to the first, they are entwined such that in space-time, electromagnetism is the only true invariant. Similar remarks apply to energy and momentum: it is 'energy-momentum' that acts as the relativistic measure of mass and motion in space-time. Thus Einstein's relativistic theory of gravity, which generalises Newton's theory, relates the density of energy-momentum to the curvature of time and space.

The essence of Einstein's general theory relates 'curvature in space-time' on one side of the equation to 'density of momentum and energy' on the other, the latter being multiplied by G and divided by c^4. In full, Einstein required ten equations of this type.

This is why. Curvature is a measure of how a line deviates from one direction to another in any of the four dimensions of space-time. To keep track of this required a separate equation for each possible combination of starting and finishing coordinates. A line going from one point, which has three spatial coordinates and a time to specify it, to another point located at some other time in three-dimensional space, gives sixteen possibilities in total. However, six of these are duplicated in the sense that curvature from x to y is the same as that from y to x (and similarly for x to or from z, y to z, and each of x, y and z with ct, where ct is the distance that light travels in time t). Thus, of the sixteen possible combinations, six are identical, which leaves ten that are independent.

In these ten equations, Einstein had found a relativistic generalisation of Newton's theory of gravity. Einstein's theory includes Newton's theory as a special case: Newton's theory corresponds to the speed of light, c, being infinite relative to the speed of the interacting bodies. For Einstein, signals travel no faster than c and there is no such thing as simultaneity; for Newton, gravitation acted instantaneously, as if c were infinitely large.

Because there are ten equations, it is difficult to make testable predictions of the theory. Solutions are known only for a limited number of circumstances. One of these concerns the orbit of Mercury, which provides empirical evidence for this warping of space-time in our solar system. If Mercury and the Sun were the only objects in the solar system, the planet's orbit would be a stable ellipse, with Mercury endlessly repeating the same circuit. However, the presence of the other planets perturbs its motion, so that its orbit slowly swings around: it undergoes what is called precession. Newton's theory predicts that this precession amounts to 532 arcseconds per century (there are 3,600 arcseconds in a degree). In reality, the precession is 43 arcseconds more than this: 575 instead of 532. This is a difference of about 8 per cent, an anomaly in the second place of decimals.

One attempt to explain this discrepancy within Newton's theory was to propose that there was a hitherto unobserved planet orbiting between Mercury and the Sun, and perturbing Mercury's orbit by the required amount. To recognise its fiery nature, this planet was named Vulcan. However, all attempts to find Vulcan failed: it does not exist.

Einstein's theory of general relativity explained the phenomenon precisely. Being the nearest planet to the Sun, Mercury feels the strongest gravitational force, moves the fastest and is most susceptible to the effects of relativity. After completing a circuit in warped space-time, Mercury does not end up in quite the same place as it would in Newton's picture. The agreement with observation is a wonderful testimony to the validity of general relativity; in particular, the energy of the gravitational field itself gives a contribution to Mercury's precession, which is absent from Newton's theory.

This was the first evidence for a gravitational field which is itself a dynamical entity. A century would pass before there was evidence for waves in the gravitational field and that the speed of these waves is consistent with that of light. This came with the first direct observation of gravitational waves, by the Laser Interferometer Gravitational-Wave Observatory (LIGO) in February 2016. A gravitational wave disturbed space-time at two LIGO detectors, which are separated by some 3,000 km. The signals were detected about 7 milliseconds apart; the wave was travelling at an angle relative to the line connecting the detectors, and when this was taken into account the speed was found to be equal to that of light, within the errors of measurement.

GRAVITY AT THE PLANCK SCALE

In Einstein's general relativity theory, gravity acts also on energy and momentum, not simply on mass. For example, massless photons of light feel the gravitational attraction of the Sun and can be deflected. Thus at a total solar eclipse, stars that are almost in the line of sight of the Sun appear displaced relative to stars that are further away from that direction, a prediction that was confirmed in 1919 and made Einstein an instant celebrity. The headline in *The Times* of London on 7 November 1919 was: 'Revolution in Science: New Theory of the Universe: Newton's Ideas Overthrown'.

Newton's rule that the force of gravitational attraction between two objects is in proportion to the product of their masses is a classical approximation which is true for objects when the effects of relativity are in the sixth place of decimals. For two particles at high energy, the force is more accurately proportional to the product of their energies, and hence larger than Newton's classical rule would predict.

This difference is far beyond our ability to detect, however. To illustrate how minute it is, recall the case of the hydrogen atom where the force of gravitational attraction between its constituent electron and proton is 10^{40} times smaller than their mutual electrical pull. Now, imagine that proton accelerated in the Large Hadron Collider to a speed at which its energy is 10^4 times greater than when almost motionless in a hydrogen atom. If it encountered an electron with a similar energy, for which the energy gain would be more like 10^7, their mutual gravitational attraction would be enhanced by nearly a trillion times relative to what Newton's theory predicts. But this would still be some 10^{28} times smaller than their electrical attraction, and thus remain utterly negligible.

Suppose, however, that we could perform a measurement on an electron and a proton deep in space, in the vicinity of some exotic object such as a black hole. Electric and magnetic fields in the cosmos can accelerate charged particles to energies far in excess of anything currently possible on Earth. Nonetheless, such events can happen 'out there', and presumably do. Consider the interaction between a proton and an electron in which each has an energy comparable to the Planck scale. This is some 10^{19} times the rest energy, or mass, of a proton, and about 10^{22} times that of an electron. Their mutual gravitational attraction, which is proportional to the product of their energies rather than their masses, will be some 10^{41} times greater than Newton would have predicted. In other words, the strength of the gravitational force will be similar to, or even greater than, their electrical attraction.

Our core theory of fundamental particles and forces can safely quarantine gravity at present, and for the foreseeable future, but this cannot be the final word. A quantum field theory that includes gravity must be included in the final theory of everything.

THE COSMOLOGICAL CONSTANT AND DARK ENERGY

But for the need to include quantum theory, general relativity could be the theory of everything to do with gravity. Suppose, then, that there were no quantum issues to cloud the scene: what questions would remain? A goal of a theory of everything is not just to describe the universe, but also to explain why it has the properties we observe it to have.

What we call gravity is a warping of the four-dimensional fabric of space-time. Einstein came to general relativity, which shows that Newton was conceptually wrong, from the principle that the laws of nature are the same in all frames of reference. His ten equations, which relate the geometry of space-time to the mass and energy within, are the simplest that satisfy this criterion; they are not, however, the most general. The principles of general relativity allow a further contribution, one which has observable consequences on cosmic scales. The so-called cosmological constant has an unusual history, linked to our changing perception of the expansion of the universe.

In 1927, the Belgian physicist and priest Georges Lemaître showed that Einstein's equations – general relativity in its simplest form, without any cosmological constant – predicted that the universe is expanding. This disagreed with the received wisdom, which was that the larger-scale structure of the universe is static in space and time. In order to make the universe stand still, Einstein added a cosmological constant to his equations. To achieve a static universe, the cosmological constant had to have some specific value. If its size were changed by a small amount, the universe would instead either contract or expand. This fine-tuning of the cosmological constant to agree with the evidence of a static universe made the theory less simple, and not as beautiful.

Two years later, in 1929, the American astronomer Edwin Hubble made a startling discovery, using what was then the world's biggest telescope. Until Hubble's observations, the known universe consisted only of our own Milky Way galaxy. Indeed, the concept of other 'galaxies' did not yet exist. Hubble showed that the universe consists of innumerable galaxies and, most remarkably, that on average they are moving apart from one another. This implied that the universe had a beginning, and an age; in modern terminology, the universe began in a Big Bang some 13.8 billion years ago.

With Hubble's discovery that the universe is not static, but expanding, Einstein was cock-a-hoop. There was no longer any need for a cosmological constant: Hubble had restored beauty to Einstein's equations. Einstein later described his incorporation of the cosmological constant as 'the greatest blunder of my life'.[4]

Would the cosmological constant have spoiled the mathematical beauty of general relativity? Einstein clearly thought so, his intuition being that the equations of his theory should be as simple as possible, so the inclusion of a cosmological constant would be an unwanted complication. However, there is another perspective on 'simplicity'. The general structure of the theory does in fact allow there to be a cosmological constant, so to remove it is to assume that its magnitude has a specific value: zero. When 'zero' appears out of an infinite number of possibilities, there is invariably some deep reason. The mass of the photon is zero, for example, due to a profound constraint on the behaviour of quantum fields known as gauge symmetry. There are other quantum symmetries, such as chiral symmetry, which require certain other phenomena to exhibit zero energy. However, there is no symmetry known that would require the cosmological constant to vanish.

Thus a cosmological constant cannot be excluded, and indeed the latest findings show that it is required. But although an empirical value for this quantity has been identified, there is no theoretical prediction of its magnitude. As far as general relativity is concerned, its magnitude is arbitrary.

A small and negative value for the cosmological constant would give a static universe, which Einstein had invoked until Hubble discovered that the universe is expanding. A vanishing value for the constant would imply a steady expansion, which for many years is what observations seemed to suggest. In the last decade, however, precision data – the sixth place of decimals again – have been obtained which show that expansion is faster now than it was billions of years ago. To accommodate this accelerating expansion, the value of the cosmological constant has to be small and positive, but not zero.

The physical significance of the cosmological constant is in effect an energy density, which fills all of space and pushes galaxies away from one another. Its origin is for now a mystery, and it is often referred to as dark energy.

Whatever its origin, dark energy appears to be consistent with general relativity theory when the cosmological constant is included.

Thus general relativity theory can accommodate dark energy, but does not explain it. One possible source could be quantum effects associated with gravity. We know that the relativistic quantum theory of the electromagnetic force, QED, accommodates virtual fluctuations of energy in the vacuum, and precision measurements of the magnetic properties of the electron, for example, verify this phenomenon. A quantum theory of gravity will also have similar implications for the nature of the vacuum. However, combining general relativity and quantum theory turns out to present deep problems, of both practice and principle.

General relativity and the quantum theory appear to have brought us to the frontier of a theory of everything. However, as I argue in the next chapter, their conjunction introduces two clouds into our twenty-first century sky. We are in a similar situation to Lord Kelvin a century ago.

6

The clouds of quantum gravity

Certain fundamental particles of matter, such as the electron, neutrinos and quarks, all have the same amount of intrinsic angular momentum, or spin. Expressed in multiples of the standard quantum amount, proportional to Planck's constant, these particles have spin 1/2. Any particle with spin 1/2, when quantified this way, is known as a fermion. Particles with integer spins (0, 1 or 2) are known collectively as bosons. Thus the Higgs boson, with no spin, is a boson, as is the photon, with spin 1.

Quantum electrodynamics, quantum flavourdynamics and quantum chromodynamics have a common feature: the carriers of the respective forces – the photon, W and Z bosons, and gluons – all have spin 1. The proof that these quantum field theories are renormalisable (in other words, that apparent nonsense such as predictions of infinite probabilities can be consistently removed) partly depends on the fact that these bosons have spin 1. There is no such happy coincidence for gravity, unfortunately.

To see why, let us first return to Maxwell's theory of electromagnetism. Electric and magnetic fields have both magnitude and direction, and are thus vectors. The electric force between two electrons has the same strength as the electric force between an electron and a proton, but acts in the opposite direction: one attracts, the other repels. This directional feature is the source of the vector, which in QED carries over to the spin 1, or vector, character of the photon. A similar bidirectional character is implicit in QFD and QCD. So what is the situation for gravity?

For Newton, gravity is universally attractive – there is no antigravity – and acts in proportion to the masses of objects. If this were the final theory, the mathematics would be simpler than in the quantum theories. However, Einstein's theory of gravity – general relativity – reveals that the source of gravitation is the relativistic combination of energy and momentum. Momentum is a vector as it can be directed in space – your motion could be in any of three dimensions: up and down, front to back, or sideways, for example. Thus when the three directions of momentum are combined with energy, there are four independent measures, and the description of the gravitational interactions involves ten linked equations, as we saw in the last chapter. In mathematical language, these equations are the ten independent components of a tensor.

A quantum field theory of a vector field involves bosons with spin 1; that for a tensor field requires spin 2. Thus in place of the electromagnetic force's massless photon, with spin 1, a quantum theory of gravity requires the existence of a massless particle, called the graviton, which has spin 2.

It is possible to calculate the gravitational interaction between two particles under the assumption that it involves the exchange of one or more gravitons. However, a complete quantum theory of the gravitational field must take into account the perturbations caused by gravitons producing virtual particles and antiparticles, as happens in QED. There is also the possibility that gravitons, which themselves carry energy and momentum, can interact with one another. If that turns out to be the case, then the possibilities multiply and their sum explodes to give infinity.

A similar phenomenon happens in QCD, where the gluons carry colour charge and can mutually interact. However, the fact that gluons are spin 1 bosons, like photons, enables QCD to be renormalised – made viable – like QED. For gravity, however, this does not happen. Gravitons have spin 2; the simplest theory of quantum gravity we could attempt to make would not be renormalisable.

If you find infinity as your answer, then you do not have an adequate theory. The problem with constructing a viable quantum theory of gravity, however, goes beyond the question of renormalisability. It goes to the very foundations of quantum mechanics and quantum field theory.

As we have seen, key to experiments based on quantum field theory is that to reveal the dynamics of physics at small distances, you need to scatter particles at high energies. However, if gravity becomes important, this founding principle is lost. To see why, imagine that we could collide two particles head-on whose energies are comparable to the Planck energy, 10^{19} GeV. The collision leads to a configuration whose total energy exceeds the Planck energy, which according to quantum theory will be localised within the Planck length, which is of the order of 10^{-35} metres. The result, according to general relativity, is that this will have created a black hole.

That would indeed be an intriguing experiment to perform. But there is nothing here that necessarily implies a fatal problem for theoretical physics. The potential paradox arises when we now imagine repeating the experiment with particles of even higher energy. What does theory predict for this?

Quantum field theory assures us that at higher energies we will be able to investigate nature on even shorter scales of distance. By Einstein's equivalence between mass and energy, $E = mc^2$, higher energy implies that the resulting black hole has a larger mass. Yet general relativity tells us that the radius of a black hole grows in proportion to its mass, so it will be bigger than before. Thus, as energy increases, the resulting black hole will gobble up more of space-time, which frustrates the correspondence between high energy and short distance inherent in quantum field theory.

This problem is exposed by the imagined experiment, but it already plagues us in practice due to quantum uncertainty. Quantum field theory implies that the vacuum of space is filled with particles and antiparticles which bubble in and out of existence on faster and faster timescales over shorter and shorter distances. We know this to be true because the calculation of the magnetic properties of the electron in QED, which is accurate to twelve places of decimals, takes into account these very phenomena.

The computation includes the effects from fluctuations on ever shorter length scales – or, equivalently, at ever higher energies. This is fine as far as QED, QFD or QCD are concerned, but gravity introduces a paradox because fluctuations near the Planck scale correspond to huge energies, comparable to the Planck energy. Black holes thus bubble in and out of existence transiently, and remove all distances shorter than 10^{-35} metres from the observable universe. Thus at very short distances, space-time itself becomes some indeterminate foam. The foundations of quantum field theory – local interactions at distinct locations in space-time – seem to have vanished. There is a fundamental conflict between quantum field theory and gravity.

Just as Lord Kelvin identified two clouds that intruded on the clear sky over his nineteenth-century physics land-scape, so in the twenty-first century our quest for a theory of everything has run into a problem with gravitational black holes. For us, there are now two clouds in plain sight whose removal will be a minimum requirement for a theory of everything. The first is the problem of the cosmological constant when we attempt to combine quantum field theory and general relativity.

THE COSMOLOGICAL CLOUD

The appearance of Planck-scale black holes undermines the foundations of quantum field theory. This is a fundamental problem: it cannot simply be put into quarantine and ignored for practical purposes because it has measurable consequences in our existing observable universe. This is why.

Our universe is expanding, and recent data on the rate of this expansion reveal that there is some repulsive dark energy in the vacuum. The application of quantum field theory to general relativity shows that the vacuum, far from being a void, will have some energy density, similar in effect to this observed dark energy. This phenomenon is accommodated in general relativity by having a finite value for the cosmological constant, which we met earlier. Unfortunately the magnitude of the observed dark energy, or of the cosmological constant, differs from what is expected in quantum field theory by an unimaginably large amount.[1]

According to quantum field theory, if the source of this energy density is an attractive force, as in conventional gravity, the universe should be curled up on the scale of 10^{-35} metres. If the contribution has the opposite sign, like some repulsive antigravity, the effect would be that the universe doubles in size every 10^{-43} seconds. This briefest of moments, known as the Planck time, is the time that light takes to travel 10^{-35} metres. Neither of these scenarios is observed in our universe. In reality, the universe doubles in size on a timescale of 10 billion years, some 10^{60} times longer than the Planck time. The mismatch in the cosmological constant is the square of this – some 120 orders of magnitude.

So, when quantum field theory is applied to the electromagnetic force, we can understand features of an electron to an accuracy of twelve decimal places, but if we try to do the same with gravity we find that we fail to describe the universe by 120 orders of magnitude. To grasp how ludicrously huge this number is, consider this: it exceeds the total number of protons in the universe by forty orders of magnitude. One desperate attempt to rescue the theory would be to suppose that there is some other contribution to vacuum energy which cancels this to a precision of 120 places of decimals. But to invent ad hoc some entity solely to rid ourselves of this quantitative disaster is not good science – it would be unnatural 'fine tuning'.

What we have here is a twenty-first-century analogue of Lord Kelvin's clouds. The problem of the cosmological constant – the vacuum energy implicit in quantum field theory – suggests that there is some fundamental flaw in our theoretical structures. I shall return to ideas about how to solve this after we meet the second of the latter-day clouds: the hierarchy problem.

THE HIERARCHY CLOUD

If the first cloud is the irony that there exists a large-scale universe at all, then the second, not unrelated cloud is that there are structures within that universe. Galaxies, stars, planets – all these objects exist because gravity is very feeble compared with the electromagnetic and strong forces. Stars survive because the radiation pressure from nuclear fusion in their heart, which is controlled by nuclear and electromagnetic forces, resists gravitational collapse. If gravity were as strong as these other forces, stars would collapse into black holes.

All the atomic elements in the periodic table, and all matter built from these seeds, exist because the spatial dimensions of nuclei and atoms are so much greater than the Planck scale. This in turn is because the mass of the electron is small. The size of a hydrogen atom is controlled by the strength of the electromagnetic force and by the mass of the electron. (If the electron were lighter than in reality, the atom would be larger.) It is because the electron is so much lighter than the Planck mass that atoms are so much larger than the Planck length of 10^{-35} metres.

The electron gets its mass by its interaction with the ubiquitous Higgs field. Experiments at the Large Hadron Collider show that the interactions with the Higgs field take place over distances of some 10^{-18} metres – for example, 125 GeV of energy is required to excite a single Higgs boson. If the Higgs mass scale were larger, or the interaction range shorter, then the mass of the electron would have been bigger than in reality.

This is where the conundrum of quantum fluctuations comes to haunt us. These fluctuations should also affect the Higgs field, and the energy required to excite a Higgs boson – the 'mass' of the boson – should become of the order of the Planck scale. Instead of the Higgs field interacting with particles on a 10^{-18} metre length scale, these dynamics should take place on a scale of some 10^{-35} metres. The masses of fundamental particles would become 17 orders of magnitude larger than we observe them to be. Atoms would be smaller by a similar factor.

As we saw for the cosmological constant, here again we have a problem of hierarchies of scale which are inconsistent with quantum gravity. The sizes of atoms and macroscopic structures only make sense if something counters the quantum fluctuations of the Higgs field. The fine tuning we would have to resort to here is not as extreme as the 120 orders of magnitude for the cosmological constant, but at more than thirty orders of magnitude it is still unnaturally huge.

These two clouds together suggest that there is some fundamental principle at work, of which we are as yet unaware. Any final theory of everything must resolve these two problems.

QUANTUM DIMENSIONS OF SPACE AND TIME

There is a conflict between general relativity and quantum field theory, at least as we presently understand them. Quantum field theory is built upon the two great pillars of twentieth-century physics: quantum mechanics, and the structure of space-time as embodied in Einstein's relativity theories. Thus to avoid these unwanted large quantum effects, which shift everything to the Planck scale, at least one of the foundations of quantum field theory must alter in some fundamental way. How convinced are we of the validity of quantum mechanics? Is our understanding of the structure of space and time beyond challenge?

In his book *Dreams of a Final Theory*, Steven Weinberg conjectured that any ultimate theory must include quantum mechanics. All attempts to construct a theory in which the rules of quantum mechanics are changed, even by a small amount, have led to logical inconsistencies. This could reflect our lack of imagination, or it may be a deep insight into the fundamental role of quantum mechanics. For the moment, at least, we can retain quantum mechanics, as there are mathematical reasons to suspect that our perception of the other pillar, the structure of space-time, is incomplete. And when we look at the maths, we find that quantum effects become welded into the fabric of space-time in a tantalising way.

The mathematical loophole goes by the name of super-symmetry. In addition to the familiar four dimensions of space and time, there are four extra dimensions of a quantum nature. Whereas it is possible to move through arbitrarily large distances in the familiar dimensions, we can only take a single step in the quantum dimensions.

In the familiar dimensions the usual rules of mathematics apply, so that for example $a \times b = b \times a$. But for variables in the quantum dimensions, things are different: $a \times b = -(b \times a)$. So the repetition of the same operation, represented by $a \times a$, vanishes because $a \times a = -(a \times a)$, which is only possible if the value of a is zero. Thus in theories built with quantum variables that satisfy supersymmetry, it is possible to take just a single step into a new dimension, or a single step out of it, but that is all.

How would such a move into a quantum dimension appear to our four-dimensional senses? When a lone particle, such as an electron, moves into the quantum dimension, it will appear to us as a novel particle. Its mass and electric charge remain the same, but its magnetic properties will be different because its spin alters. In general, what appears as a fermion in our space-time becomes a boson, whereas a boson coverts into a fermion. The supersymmetric version of the electron is called a super-electron, usually abbreviated to selectron.

In supersymmetry theories, the large quantum fluctuations produced by conventional particles, such as the electron, are matched by corresponding fluctuations associated with their supersymmetric partners. The beauty is that these supersymmetric fluctuations have the same magnitude as for conventional particles, but they are negative. The result is that in the grand summation, the contributions from conventional particles and their supersymmetric counterparts cancel to nothing.

To a theoretical physicist, the mathematical forms in supersymmetry have a precision comparable to the music of Bach, and a depth comparable to Beethoven's ninth symphony. The algebraic structures of quantum fluctuations in space-time and those of their supersymmetric counterparts appear quite different at the start, much as the symbols on a music score for individual voices in a choir reveal nothing special. However, when combined together by skilled performers, a wonderful ode to joy results, both for Beethoven and for the mathematician.

But this analogy is incomplete. Works of art are judged on their intrinsic quality; constructions in theoretical physics may be beautiful, but if nature does not make use of them they are irrelevant. This is the crux of physics: experiment decides which ideas survive or die. Supersymmetry is a beautiful theory, but nature does not appear to read its equations in a perfect symmetric form. In a nutshell, there is no selectron with the same mass as an electron; there is no supersymmetric analogue at the same mass as that of any other fundamental particle.

There is one loophole, however. We know of other examples in physics where symmetry is present in all but the masses of particles, such as with the massive W or Z boson of the weak force in contrast to the massless photon of the electromagnetic force. Mass could hide an otherwise perfect supersymmetry, such that the selectron is much heavier than an electron, and likewise other supersymmetric particles could be much heavier than their conventional analogues, while all other features of supersymmetry remain perfect. If nature were supersymmetric at distances less than 10^{-18} metres, such that supersymmetric particles occur in the region of 100 GeV to 1,000 GeV (1 TeV), this would still remove the huge fluctuations in quantum field theory and solve the hierarchy problem. But this hypothesis can only survive if supersymmetric particles are found to be produced in experiments at the Large Hadron Collider, which is able to probe these energies and scales of distance.

As of 2017, there is no direct evidence that nature reads supersymmetry. There have been some indirect hints, in that supersymmetry theory might require the existence of relatively massive dark particles – ones which feel only the gravitational and possibly the weak interaction, but not the electromagnetic or strong forces. The motion of stars in remote galaxies, and also the gravitational response of one galaxy to another, can only be understood if the source of gravitational force is much more extensive than we can account for by observations made at optical, radio, infrared, X-ray and gamma-ray wavelengths. The known particles which make up the observable universe seem to be mere flotsam on a sea of so-called dark matter, stuff that does not shine at any electromagnetic wavelength. What this consists of we have yet to discover. However, if it turned out to consist of dark particles, in accordance with supersymmetry theories, that would make for a remarkable symbiosis of cosmology, particle physics and our understanding of space-time. It would also remove one of the two twenty-first-century clouds.

It may also enable us to take a big step towards the resolution of the other puzzle, of the cosmological constant. To see how, we first need to introduce the idea of superstrings and the multiverse.

SUPERSTRINGS

Superstring theory is built on a fundamentally novel conception of the nature of space-time at Planck scales of distance, and potentially it also has implications for the nature of the cosmos at scales of many billions of light years.

In string theory,[2] space-time contains glitches on a scale of 10^{-35} metres. These slits in the smooth fabric of space-time are known as strings, which can vibrate in an infinite numbers of ways. They are much too small to be detected in our current experiments, which show space-time appears to be continuous. The hypothesis is that, at large distances or at energies far below the Planck energy, quantum field theory is an approximation to an underlying string theory. The fields we currently associate with particles are, at source, vibrations of these slits in space-time.

Initially, in the late 1960s, string theory was a mathematical exploration. It suggested that there was a mode of vibration which acts like a particle with no mass, and with spin twice that of a photon. This happens to be identical · to the graviton, the hypothesised quantum bundle of the gravitational field. Just as the exchange of photons drives the electromagnetic force, so the force of gravity is down to the exchange of gravitons. The feeble strength of the gravitational force sets the scale for the dynamics of the strings. If this specific mode is indeed to be identified as the graviton, then the weakness of gravity implies that the string must be very hard to excite: its tension is huge, and the energy needed to excite it has to be of the order of the Planck scale.

The implication is that all known material particles are manifestations of the lowest modes of the strings, analogous to an open string on a violin. Conversely, any excited mode, analogous to a harmonic on a violin string, will materialise in the vicinity of the Planck energy. That would make direct experimental proof of string theory hard to obtain. We have to hope that there are theoretical consequences of the theory which we can test at the energies within our reach. Definitive evidence is not yet available, but if we accept the premise that the ultimate foundations of reality are to be found in strings, what might we conclude?

First, quantum field theory is probably not fundamental, but rather a general property of any theory that satisfies the constraints of quantum mechanics and special relativity. Thus a viable theory will appear to be a quantum field theory at low energies, where 'low' is relative to the Planck scale. Conversely, the fact that nature is well described by quantum field theory at presently accessible energies does not imply that this remains true at the Planck scale. Our present core theory of particles and forces, known as the standard model, might thus be a low-energy illusion of a richer theory whose identity will only become clear at higher energies. This philosophy is consistent with the history of physical science: Newton's classical mechanics is a low-energy approximation to Einstein's special relativity, and Newtonian gravity is a low-energy illusion of general relativity. I shall return later to the relation between classical and quantum mechanics. For now, it is sufficient to say that this will complete the recipe for our conception of a theory of everything.

There were a huge number of possibilities for string theories, but in 1984 the British theorist Michael Green and the American theorist John Schwarz discovered that only two of them appeared to be consistent with requirements of quantum field theory at low energies. One important piece of their analysis was the incorporation of supersymmetry into the mix: superstring theory was born. The slits in space-time included six further dimensions in addition to the familiar four. Suddenly there was hope that superstring theory might be so tightly constrained that a unique theory of everything was at hand.

There followed a vast amount of theoretical investigation into the foundations of superstring theory. This has had enormous success in developing new mathematical insights, and there have been applications in unexpected areas of physics, such as quantum entanglement and information theory. Ironically, however, there has been little obvious progress with the original quest to make super-strings the unique theory of everything about particles and forces.

Much of the hype of the final decades of the twentieth century cooled when a vast number of superstring theories were formulated, each of which was consistent with the quantum mechanical constraints that Green and Schwarz had identified. The original hope for a unique theory has gone. However, there remains great confidence that string theory has unveiled an essential truth about the fissured structure of space-time at the Planck scale. In particular, at low energies superstring theory appears as a quantum field theory, but without the major problems of the hierarchy of scales and inconsistency at the Planck scale that plagued applications of traditional quantum field theory.

There is one generic feature, though, which could be within the reach of experiment: superstring theory requires that supersymmetry is realised in nature. Thus super-string theory could potentially remove one of the clouds of quantum gravity, and so there is an intense experimental effort to find direct evidence for supersymmetry, such as the selectron or other predicted supersymmetric analogues of the known particles.

Once advertised as the long-sought unique description of our universe, superstring theory has turned out to be like the mythological Hydra of Lerna, which grew two heads for each one that was cut off. With the hope for a unique theory now gone, the discovery of a vast variety of mathematically consistent versions of superstring theory has generated mixed reactions.

For some, superstring theory is no more than 'a beauti-ful set of ideas that will always remain just out of reach'.[3] Others think that the emergence of a multitude of super-string theories might provide a way to remove the second cloud, by solving the enigma of the cosmological constant.

In this interpretation of its equations, superstring theory has evolved such that our universe is no longer 'Everything'. Instead, we are in what is just one of a multitude of universes: the multiverse, in which all possible values of fundamental parameters occur. In this richer 'Everything', we are the lucky winners, inhabitants of a Goldilocks universe suitable for life.

THE MULTIVERSE

Superstring theory, which describes space-time at unimaginably small scales, has shown that we might have to change our concept of space-time at cosmic scales. It turns out that a unique and specific superstring theory of space-time at the Planck scale will generate a multitude of different solutions for the large-scale properties of space-time – a multitude of universes. The basic laws of nature are the same in all these universes. For example, they all feature a force between entities in proportion to their energies, hence gravity. There are particles with electric and colour charges, capable of changing flavour, hence electromagnetic forces and analogues of the 'weak' and 'strong' forces. (I have used quote marks here because the relative strengths of the weak and strong forces will in general differ from their values in the particular solution that corresponds to our universe.)

It is estimated that there are 10^{500} solutions in which supersymmetry is manifested in ways that roughly correspond to our situation. This number is so enormous as to be incomprehensible: compared with 10^{120}, the scale of the cosmological constant mismatch, it is mind-bogglingly vast. In this gargantuan number of possibilities, magnitudes for the cosmological constant will range from huge positive values down to zero and then all the way to huge negative values.

A universe with large negative values of the cosmological constant would grow and collapse in too short a time for sentient life to evolve. At the other extreme, large positive values imply a ubiquitous repulsive force which would cause a universe to expand so rapidly that the attractive force of gravity would be overwhelmed. In such a universe matter would not cluster into galaxies of stars, which are prerequisites for life-supporting planets. In order for intelligent life to arise, it would seem that the cosmological constant – with all the quantum fluctuations taken into account – has to be small enough for there to be a good chance of galaxies forming and stars living for billions of years. This is obviously true for our universe – otherwise we would not be here to debate the question. That may be true, but it does not explain why we won the great lottery. There is, however, an interesting hypothesis in cosmology, known as eternal inflation, which has the potential to solve this problem.

Observations of the rate at which the universe is expanding, and the fact that the cosmos appears to be remarkably uniform throughout an observable range of up to some 27 billion light years, suggest that there was a brief period of rapid expansion, termed inflation, during the Big Bang. One consequence of this is that the universe extends much further than we can ever see. An analogy is that on the surface of the Earth our view is limited to within the horizon, even though there is much that extends beyond. Thus the horizon for our universe is some 13.8 billion light years away, beyond which there are more remote galaxies which are inaccessible to us.

Models of this inflation, which agree with observations within the accessible universe, suggest that bubbles of other potential universes are forming continuously. The values of the cosmological constant in these bubbles are randomly distributed across an infinite spectrum. There is an infinite multiverse where all 10^{500} possible solutions of the superstring dynamics are produced, somewhere in space-time, infinitely often.

Thus nature is manufacturing universes continuously. It is then guaranteed that one (at least!) of these 10^{500} possibilities happens to have a cosmological constant of a small size, tuned at a precision of one part in 10^{120} like ours. Superstring theory, originally thought to be a unique guide to the raison d'être for our one and only universe, happens to have a gargantuan number of solutions, but far from being an insurmountable problem this is now seen as a godsend. Superstring theory is currently the only explanation we have for the smallness of the cosmological constant: we happen to live in a Goldilocks universe, which is part of a multiverse.

EVERYTHING OR NOTHING?

To have any chance of eliminating the two clouds in our modern sky, we need a successful marriage of quantum field theory and general relativity. To summarise the problem: increasing energy in quantum field theory probes shorter distances; black holes grow with energy and so obscure larger distances. It seems therefore that a viable theory of quantum gravity cannot be a traditional quantum field theory of the form we have used to describe the other fundamental forces.

It is possible that the final theory of everything, which includes quantum gravity, will have novel features in general but will reduce to quantum field theory at low energies. As we have seen, this would be in accord with history: Newton's mechanics is a low-energy limit of Einstein's special relativity; Newtonian gravity is an approximation to general relativity. Quantum field theory may therefore also be a low-energy limit of some richer theory. Mathematical examples of such a theory include string theory and loop quantum gravity. Whether either of these is a guide to the final theory, however, remains an open question, and one that may prove impossible for experiment to answer. This raises profound questions about the nature of science and our quest to establish a theory of everything.

To be specific, let us suppose that we have already found the key (if not yet unlocked the complete theory) in the insight that space-time contains quantum dimensions, which incorporate supersymmetry, and that superstring theory encodes the deepest dynamics of nature. What are we to conclude about its implication that we inhabit a multiverse? Scientific knowledge is supposed to be empirical: to be accepted as scientific, a theory must be falsifiable, at least in principle. This argument was advanced in 1934 by the philosopher Karl Popper, and is generally accepted by most scientists today as determining what is and is not a scientific theory. Are superstring theory and the idea of the multiverse – the existence of multiple universes – falsifiable?

The true dynamics of quantum gravity would become apparent in experiments at the Planck energy scale. This is a thousand trillion times greater than the highest energies accessible at the Large Hadron Collider, however. If we were in a position to plan experiments under such conditions, we could no longer ignore quantum gravity, but because this region is so remote it can be quarantined, for now, in our core theory, the standard model. That is the good news, making possible practical science and technology today in the absence of an agreed theory of quantum gravity. The downside, however, is that this very remoteness hides any clues about how to construct a viable theory of everything.

At present, string theory faces a hurdle that previous theories of everything have all overcome. It is this: the theories of everything we have met so far have used a few assumptions, and been able to explain a lot about the world, admittedly in restricted regimes. This 'a lot for a little' has made them impressive and important. From that perspective, string theory – as a candidate that unifies the laws governing the behaviour of particles and forces – has stalled: it is 'an exploration of fascinating mathematical structures that may or may not relate to the physical universe.'[4]

String theory at least passes Popper's falsifiability test and so qualifies as a scientific theory. It is not intrinsically untestable, but there has been no success yet. In experimental terms, one can imagine some future technology that is, in theory at least, capable of accelerating particles to the Planck scale. In the foreseeable future, the challenge is whether evidence to support string theory can ever be found through experiment in a real world in which particle accelerators capable of reaching the Planck energy scale will remain beyond our capability.

One line of investigation concerns a central plank of the theory: the property known as supersymmetry. The discovery of supersymmetric particles, though it would not be evidence uniquely for string theory, would be a significant step. Without doubt, the discovery of supersymmetry would take us beyond the present standard model and require a new theory of everything. And string theorists will, with justification, point out that, mathematically, there is already a candidate for that theory of everything: string theory.

We can therefore hope that evidence for supersymmetry may be found in the not-too-distant future. When superstring theory is dressed with the resulting information about supersymmetric particles, it should be possible to see whether the resulting quantum theory of general relativity solves the hierarchy problem, namely that the masses of the Higgs boson and other particles are stable at their observed values and are not sucked off to the Planck scale by the forces of quantum gravity.

However, the question of the cosmological constant would still remain. It seems inconceivable that a fine tuning of 120 orders of magnitude could be explained by the discovery of a menu of supersymmetric particles at the Large Hadron Collider. More likely, it seems to me, is that the multiverse aspect of superstring theory would become the prime suspect for an explanation of the cosmological constant problem.

However, it is more problematic to establish the multiverse scientifically. The reason is immediate: as there is no possibility of communication between us and other universes, there is no empirical way to test the multiverse theory. This raises a profound question: if a scientific theory is elegant, and is consistent with known facts, does it need to be tested by experiment? Could a theory of everything be acceptable, based on this criterion, even if it cannot be solved? This question has come to prominence in recent years, especially following the discovery of the multiverse solutions to superstring theory, which are beyond the reach of experimental falsification, even in principle. The uncomfortable question that arises is whether this would still be regarded as science, and whether such a candidate theory of everything is, from a scientific perspective, a theory of nothing.

Many scientists regard the very concept as unscientific. This 'kaleidoscopic multiverse comprising a myriad of universes', in the words of George Ellis and Joe Silk,[5] sets a basic challenge: the suggestion that another universe need not have the same fundamental constants of nature as ours raises the question of just what determines the values in our universe. 'In a general multiverse model, everything that can happen will happen somewhere, so any data whatever can be accommodated. Hence it cannot be disproved by any observational test at all,' says Ellis.[6] If we agree with Popper that a theory must be falsifiable to be scientific, then by implication, the multiverse concept lies outside science.

The existence of the multiverse is unlikely to be confirmed by observations in our specific 'sub-universe'. But Steven Weinberg argues that this is not necessarily fatal to the theory's scientific validity. 'The multiverse idea is very speculative,' he says, 'but it's not an entirely unreasonable speculation. The existence of a multiverse might some day be confirmed by deducing it from a theory that [has been verified] by the success of sufficient other predictions.'[7]

Scientific theories can still be of use even when they are only partially understood. One example has permeated much of our narrative: quantum theory. This is probably the foundation of physics most likely to survive in a final theory, yet is full of concepts that appear to contradict our intuitive notion of how things behave. The theory of quantum mechanics is science because it can in principle be disproved. It has survived innumerable tests and made countless successful predictions.

The multiverse may be a productive mathematical device, but that does not require all its component universes to have 'reality'. We might heed the warning of the German mathematician David Hilbert: 'Although infinity is needed to complete mathematics,' as Hilbert is reputed to have said, 'it occurs nowhere in the physical universe'. This underlies the infinity puzzle of QED, for example, which threatened to kill the theory until the techniques of renormalisation enabled this unphysical piece of mathematics to be eliminated.

This is the crux. Mathematical concepts are tools which enable us to investigate reality, but themselves do not necessarily imply physical reality. Beauty may be a guide in pursuit of a theory of everything, but it cannot be the judge: evidence in support of a theory has to be experimental or observational, not simply theoretical. We should let history be our guide, as experiments have proved many beautiful and simple theories wrong.

For example, in the seventeenth century the German astronomer Johannes Kepler became convinced that he had discovered the explanation for the structure of the solar system. He took each of the five regular polyhedrons (tetra-hedron, cube, octahedron, dodecahedron and icosahedron) and fitted a sphere inside and outside each of them. When he nested these spheres within one another, he found that the radii of the spheres were – almost – in proportion to the radii of the orbits of the six planets then known. His 'theory' had a seductive geometrical beauty, which convinced him that he had stumbled on God's plan. He wrote: 'I feel carried away and possessed by an unutterable rapture over the divine spectacle of heavenly harmony.'[8] But his theory was false: Kepler's planetary model was eventually undermined, not least by the discovery of further planets.

Although Kepler was wrong in his description of the arrangement of the planets, the emergence of better data led him to an accurate description of their motion: planetary orbits are not circles, but ellipses, and the Sun is located not at the centre but at one focus of the ellipse. These insights subsequently inspired Isaac Newton to his law of gravity, and the start of a four-centuries-long quest for a theory of everything.

QUANTUM THEORY OF EVERYTHING

Whereas snooker balls bounce off one another in a way determined by Newton's laws, beams of atoms, which obey the rules of quantum mechanics, scatter in some directions more than others. Newton's dynamics are deterministic, whereas quantum mechanics is not, and that is a profound difference. The belief that classical mechanics would be fundamental to a theory of everything seems undermined by this lack of determinism in the passage to quantum mechanics. If quantum mechanics is an essential plank of the final theory, how can Newton's laws claim to be fundamental?

The answer is that neither determinism nor Newton's mechanics is fundamental. Each is an effective description that emerges from the basic quantum rules. There have been clues that classical mechanics is not fundamentally deterministic since the eighteenth century, though their significance seems only to have been fully recognised in recent decades. The key is a formulation of classical mechanics discovered by the French mathematician and astronomer Joseph-Louis Lagrange in the century after Newton, which reveals that determinism in classical mechanics is not fundamental but instead is an example of an 'organised behaviour emerging out of simple rules'.[9]

To appreciate how this arises, let us return to the basic challenge of classical mechanics: if you know where some objects are now, where will they be at some future moment? Newton's laws of motion, which allow us to answer this, inspired the concept of energy. There is the energy associated with motion, kinetic energy, and also potential energy, where the position or state of an object gives it the potential to gain kinetic energy. The sum of the potential and kinetic energies remains constant.

Lagrange's genius was to focus on the difference between a body's kinetic energy and its potential energy. If you add up the magnitudes of this difference at every point on the body's trajectory, from beginning to end, you obtain a sum (or integral) known as the action, which has the same dimensions – the product of energy and time – as we met earlier. The remarkable feature is that the path taken by a body to get from one point to another in a specified amount of time is the one with the least action. In general, the paths that minimise the action are the same as those that satisfy Newton's laws.

For example, the natural tendency of a body, free of external forces, to travel in a straight line, rather than on any of an infinite number of possible zigzags or curves, is because the shortest path in such a case has the least action. In the quantum world, though, particles seem able to go anywhere, deviating from a straight line even when no forces are acting on them. How, then, does the principle of least action emerge, and validate the classical mechanics of Newton?

Lagrange's focus on the action reveals an unexpected mystery in classical mechanics. It is as if, before setting off, a body first samples all possible routes, calculates their actions and then decides on the magic solution. This purposeful aspect of action, whereby a body in classical mechanics seems to know beforehand how to get to where it wants to be, is actually rather eerie.

In 1946, by focusing on the action, the American theorist Richard Feynman showed how classical mechanics emerges from the underlying fundamental quantum mechanics. He began by assuming that all paths are possible, not just those with the least action. He imagined time sliced into pieces and asked, if a particle is at some point at time zero, what is the probability of it being at some other point at a specific later time? In his formulation, the probability is the square of a complex number known as the probability amplitude, which is simply related to the action. The magnitude of this amplitude oscillates along any path like a wave.

The idea here is first to calculate the value of the action for each path the particle could take to reach the other point, including trajectories that are absurd in normal experience. When a group of particles is gathered together to form a large object, such as a molecule, their individual amplitudes mutually cancel out for all paths except those that are very near the classical one.

These ideas may seem strange, but they are actually rather familiar: they parallel how the ordered geometry of light rays emerges from spreading undulating waves of electric and magnetic fields, which radiate from a source in all directions. The French mathematician Pierre de Fermat discovered the golden rule in the seventeenth century: out of all possible paths that waves of light may take between one point and another, the actual path is the one for which the light takes the least time. Along this route they appear as simple rays.

An analogous approach applies to atomic particles, such as the electron. In Feynman's vision, nature is utterly democratic, placing no constraint on where an electron can go. An electron samples all possible paths in both space and time. The waves of values in Feynman's probability amplitude for the electron's various possible paths would mutually self-destruct everywhere but for the shortest 'optical' path, thereby giving an appearance of travelling in rays, as particles do.

Feynman showed that classical mechanics emerges from quantum laws, and the link is the action. Action has the same dimensions as Planck's constant, h. Quantum mechanics rules when the action is of similar size to h or smaller. When the action is much larger than h, however, quantum democracy gives way to the more rigid phenomena that we term classical mechanics. Thus, classical mechanics emerges from the fundamental quantum mechanics when the action is large compared with Planck's h.

Quantum mechanics is not deterministic, but neither is classical mechanics at a fundamental level. Classical mechanics only appeared to be so because of the way Newton formulated his theory, and because of Einstein's later generalisation of Newton's construct. In reality, determinism is a derived notion, and Lagrange's focus on action is nearer to nature's quintessential axioms.

The lesson here is that what appear to be fundamental truths about nature might well be practical truths, but are not necessarily fundamental. Lagrange's approach to classical mechanics revealed that Newton's laws emerge from deeper levels of reality. The quantum world is immersed in this very layer.

Thus quantum mechanics may indeed be fundamental

to a final theory, as Weinberg has argued. How much else of what we regard as fundamental will be there? Quantum field theory is an effective theory, based on space-time locality. As the Princeton theorist Nima Arkani-Hamed has argued, 'There must be a new way of thinking about quantum field theories, in which space-time locality is not the star of the show.' He then suggests that, 'Finding this reformulation might be analogous to discovering the least-action formulation of classical physics. By removing space-time from its primary place in our description of standard physics, we may be in a better position to make the leap to the next theory, where space-time ceases to exist.'[10]

We have seen examples of how phenomena and concepts emerge for large numbers of particles which were not apparent in the underlying fundamental theory. For example, Newton's equations say nothing about the direction of time, yet they underpin phenomena from which the second law of thermodynamics and time's arrow emerge. The dynamics of a single snooker ball, meanwhile, respects the time reversibility of Newton's laws. The ball is made of a large number of cooperating particles whose combined action vastly exceeds the size of Planck's h, and the classical results emerge.

For Feynman, the paths of individual electrons pass forwards and backwards in time as if time itself has no fundamental presence. Bizarre as this may appear, the existence of antimatter is one of its implications. Newton's theory of everything mechanical emerges when large numbers of atoms cooperate, and thermodynamics, with its arrow of time, appears when large numbers of macroscopic constituents are involved. But at the deepest level, of a single atomic particle, when viewed from Feynman's perspective the notions of space and time themselves become hazy.

A 'sense' of time is itself a product of our macro-senses, which are a consequence of a large number of atoms cooperating. Phenomena and concepts arise for macroscopic systems which are not at all apparent in the fundamental theory when it is applied to just one or two particles. One carbon atom may be identical to another, but put enough of them together and they can become self-aware.

This brings us back to the mantra we met in Chapter 2: life, the universe and everything. Life, certainly, emerges, as perhaps the universe does too. As for the underlying fundamental theory of everything, that may still emerge – given time.

7

Back to the future

If Lord Kelvin gave his 'two clouds' speech today, he would surely identify attempts to determine a viable quantum theory of gravity as a blot on the landscape. There appears to be a fundamental problem with the union of quantum field theory and quantum theory of gravity, which for some scientists is a hint that space and time might emerge from some more fundamental concept. We have seen that quantum mechanics and/or our picture of space-time must be naive. The thought experiment about collisions at energies above the Planck scale, which produce black holes and block our ability to resolve shorter length scales, gives us a clue to the richer nature of the successor theory.

Quantum mechanics is founded on Heisenberg's uncertainty principle, according to which the spatial resolution, x, probed by an ultra-high-energy particle of energy E is indeterminate, or 'uncertain', by an amount given by of the order of the velocity of light multiplied by the ratio $\hbar/E$, where $\hbar$ is Planck's constant divided by 2π. This relation between distance and energy is the lower curve in Figure 9. The growth in size of a black hole with energy limits the ultimate resolution, however. The 'forbidden' region is shaded in the figure. At energies which are low relative to the Planck energy, this creates no practical limitation and the traditional uncertainty relation – which implies that as energy approaches infinity, the resolution of distance will tend to zero – holds. At very high energies, however, the forbidden region overwrites the naive quantum uncertainty.

The effective combination of these factors is illustrated by the solid curve with a distorted U-shape in Figure 9. The implication is that there is a minimum length accessible to experiment. It is a matter of choice whether you prefer to picture this as a modification to space-time – namely, a granularity introduced by the marriage of traditional quantum

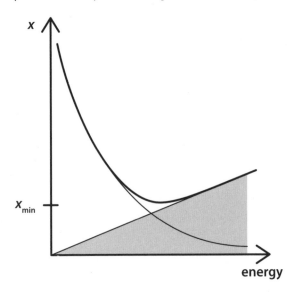

Figure 9: Black holes and quantum uncertainty. The quantum uncertainty between distance and energy eventually enters a region (shaded), which is inaccessible to experiment due to the creation of black holes. The resulting correlation between distance and energy becomes the U-shaped curve. The value of x_{min} is about 10^{-35} metres when the energy is 10^{19} GeV (gigaelectronvolts).

theory and general relativity – or as an illustration that our present description of quantum uncertainty, as expressed by the lower curve, is merely a low-energy approximation to a richer relation, described by the total U-curve.

Whichever you prefer, there is a historical precedent. As Newton's laws are a low-energy approximation to Einstein's richer theory of relativity, so is quantum mechanics, based on Werner Heisenberg's form of quantum uncertainty, a low-energy limit of a more complete theory founded on the uncertainty expressed in the U-curve.

The enormous energies that we need to harness if we are to reveal the dynamics of quantum gravity, where the notion of space and time evaporate, seem to be out of reach for any foreseeable future. There may be hope of learning about these conditions, however, because the experiment has already been done – some 13.8 billion years ago in the Big Bang. The quantum fluctuations that bubbled forth during that epoch have left their imprint on the cosmic microwave background, which is the oldest light in the sky. When first discovered, in 1964, it appeared as a uniform spectrum of electromagnetic radiation, in thermal equilibrium with a temperature of about 3 degrees above absolute zero. Half a century later, the data reveal variations of temperature in the fifth place of decimals. These trace fluctuations in the density of matter in the early universe, and are an imprint from shortly after the Big Bang. In theory, they could be decoded to reveal the dynamics of quantum gravity and whether there is a sign of the role of superstrings in those first moments. There is a chance that such questions could be answered in the coming decades.

The generally accepted theory that our universe expanded from an initial Big Bang does have some problems. Our universe appears to be large, smooth and broadly the same in all directions. What we see 13 billion years beyond the north pole looks broadly the same as what we see 13 billion years beyond the south pole. This implies that these two regions of the universe have at some time in the past been connected. If the universe has been gently expanding for its entire history, however, these two regions, which are over 25 billion light years apart, can never have communicated.

The implication, according to modern theory, is that immediately before the present gentle expansion there was an extremely brief period of enormous cosmic inflation. In less than 10^{-33} seconds, the universe ballooned from a quantum speck to over 10^{78} times its initial volume. After this initial burst of inflation a more gentle expansion followed.

Why this inflation happened, what occurred before it (if this question is even meaningful) and how the colossal inflation was driven, and then dissipated, are three fundamental questions. So even if cosmic inflation is fundamental to the final theory, with details still to be worked out, the arrival of a final theory of everything will not herald the end of physics because this trio of questions remains.

According to superstring theory, the fundamental fabric of the newborn universe contained more than the four dimensions that we know as space and time. Whereas the latter have grown with the universe over 13.8 billion years, the other dimensions remained only on minuscule scales of the order of the Planck length. What we perceive as point particles would, if resolved at the scale of 10^{-35} metres, be revealed as one-dimensional slits – strings – and higher-dimensional entities called branes (as in 'membrane'). These structures all exist in a ten-dimensional vista.

During inflation, quantum fluctuations in space-time cause the rate of inflation to vary at different places and moments. Like points on the surface of an expanding balloon, which jiggle as the balloon grows, some regions of space-time will be marginally closer together while others will be slightly farther apart. So regions of increased density matched by relative voids will emerge. These jiggles in space-time correspond by quantum uncertainty to the bubbling in and out of existence of particles. The energy driving inflation in our four large-scale dimensions can dissipate in the other minuscule ones. When the energy driving inflation is spent, the gentler expansion remains. The local modulations in energy are like gravitational wells, into which matter accumulates over the eons to form clusters of galaxies. Thus fluctuations in the cosmic microwave background are the primeval imbalances that seed our present structured universe.

Although it is unlikely that technology will grant us direct access to such conditions in our lifetimes, we are fortunate that nature has already performed the experiment – if theory is correct – during the inflationary epoch of the early universe. The results are imprinted in the cosmic microwave background as fluctuations in its temperature at a level of one part in a hundred thousand, reminding us of Albert Michelson's remark about looking for future truths in the sixth (or here, the fifth) place of decimals. Nature may also repeat aspects of these conditions when black holes collide and fuse, for example.

The challenge, therefore, is to be able to identify the traces from these events, and to decode them. In 2016, the first detection of gravitational waves was announced. These waves were generated when two black holes collided. The results confirm general relativity, and that gravitational waves spread at the speed of light. The existing pair of Laser Interferometer Gravitational-Wave Observatory detectors should in the next few years be joined by a third one, based in India, enabling gravitational waves to be observed from three slightly different angles. This should (hopefully) establish that gravitational waves behave as if generated by a tensor field – in contrast to the vector fields associated with the photon or W boson, for example.

The era of cosmic inflation is encoded in the cosmic microwave background. If we can decode its patterns, then we can in principle deduce the nature of the dynamics that apply at the Planck scale. The cosmic microwave background radiation is expected to be polarised, just as scattered light is (generating an industry in polarised sunglasses). By measuring its polarisation we can hope to learn how gravitational waves were created from the initial quantum state.

We are thus in a situation now similar to that in which nineteenth-century scientists found themselves when they were making measurements of atomic spectra. They found patterns of lines, like a bar code, but did not understand their significance. These spectral lines were later decoded to reveal evidence of atomic structure, leading eventually to the removal of Kelvin's clouds. Fluctuations in the cosmic microwave background are our twenty-first-century analogue of nature's bar code. If we can interpret their message, we may be able to resolve the dynamics of quantum gravity and remove the clouds that presently obscure the theory of everything.

What might we expect of the next theory of everything? If anyone knew the answer, they would be on the shortlist for a Nobel Prize. In the meantime, in the absence of an answer, we can summarise the possible 'quarantines' that might yet lurk in our current best theories.

One is the modification of quantum uncertainty, which we saw earlier. Quantum mechanics may be part of the final theory, but the uncertainty principle upon which it is founded – that high energy accesses shorter intervals of space and time without limit – is an approximation to some more general rule. There are already clues in the formation of black holes that at very high energies an increasing extent of space-time is inaccessible to experiment, as illustrated in Figure 9. If this is indeed how nature works, it shows that our present quantum theory is a low-energy approximation to a richer structure. It seems likely that the final theory of everything, if there is one, will be built on an extension of the correlation between energy and space-time. Our existing quantum mechanics, built on Heisenberg's uncertainty relation, is thus an approximation which is valid only as long as the energy is significantly below the Planck scale of 10^{19} GeV.

You may be thinking, 'Here we go again': what appears to

be a theory of everything turns out to be an approximation which remains valid only so long as some quantity, which is relevant to a richer theory, is too remote to be significant. We have met speeds that are trifling relative to the speed of light, and action whose magnitude is large relative to Planck's constant, being the conditions for Newton's mechanics to qualify. Gravity was ignored completely in special relativity, while general relativity appears to be valid in practice as long as quantum effects are neglected. There is also a possibility that general relativity is only true for feeble gravitational fields, and that corrections or extensions are necessary for strong gravitational forces. This could be tested experimentally in the next decade as the results from gravitational wave detectors accumulate. If strong gravitational fields reveal new truths, they may impact on our understanding of the Big Bang, and perhaps also illuminate an outstanding puzzle: was the Big Bang really the start of time, or was there something that preceded it?

The arrow of time, which we have seen emerge from mechanics when large numbers of particles are involved, links to an increase in disorder, or entropy. According to thermodynamics, a state of minimum entropy, or maximal order, requires temperatures that approach absolute zero. This leads to a conundrum: why was the universe in a low-entropy, highly ordered state, at the Big Bang? Is this a clue to what existed 'before' the Big Bang? I have put quotes around 'before' because if time – at least, as we understand it – began at the Big Bang, there is no easy definition of 'before'. So my conjecture is that in some future theory of everything, space and time will turn out not to be fundamental and will emerge from some deeper concept. Whoever first establishes what this is will enter the pantheon of science, along with Newton, Maxwell and Einstein. For now, the quest for the ultimate theory of everything goes on.

Notes and references

CHAPTER 1

1. Speech at the Dedication Ceremony for the Ryerson Laboratory of the University of Chicago, 1894. Quoted in the Annual Register of University of Chicago, 1896, p. 159.
2. Folklore attributes this remark to Lord Kelvin, but there seems no evidence that he actually said it in 1900. It does not appear in his 'two clouds' speech of that year. He may have said these words earlier, and inspired Albert Michelson: see e.g. Steven Weinberg, *Dreams of a Final Theory*, New York, Vintage, 1994, p. 13; and also en.wikiquote.org/wiki/William_Thomson#cite_note-1.
3. Lord Kelvin, 'Nineteenth Century Clouds over the Dynamical Theory of Heat and Light', *Notes and Proceedings of the Royal Institution*, vol. 16 , 1902, p. 363. The speech was given on 27 April 1900.

CHAPTER 3

1. Daniel Bernoulli, *Hydrodynamica*, 1738, as quoted by J. Bernstein, *Secrets of the Old One: Einstein 1905*, New York, Copernicus Books, 2006, p. 106.
2. The electronvolt (eV) is a unit of energy: it is the amount of energy an electron gains when accelerated by a potential difference of one volt. Thousands, millions and billions of eV are kilovolts, megavolts and gigavolts, denoted by keV, MeV and GeV. At room temperature, the kinetic energy of an atom in the air is typically about 1/40 eV.

CHAPTER 4

1. The photoelectric effect is often described, erroneously, as the proof that the photon is a particle. In fact, it is only half the proof. The photoelectric effect demonstrates the transfer of energy, but the proof that a particle

is involved also requires it to be demonstrated that momentum is conserved. Arthur Compton won the Nobel Prize in Physics in 1927 for his demonstration that energy and momentum are both conserved when light scatters from a charged particle such as an electron. The eponymous Compton scattering is thus the complete proof of the particulate nature of the photon.

2. Angular momentum is a property of rotary motion analogous to the more familiar concept of momentum in linear motion.

3. Lord Kelvin gave his 'two clouds' speech on 27 April 1900. Max Planck had offered his first theory of blackbody radiation in 1899, but experiment soon disproved it – hence one of Lord Kelvin's clouds. In the latter half of 1900, Planck made several attempts to modify his theory. The first mention of energy quantisation appears to have been in his presentation to the German Physical Society on 14 December 1900 (as quoted in www-history.mcs.st-and.ac.uk/Biographies/Planck.html).

4. Paul Dirac, 'Quantum Mechanics of Many-Electron Systems', *Proceedings of the Royal Society A*, vol. 123, 1929, p. 714.

5. The product of c and $\hbar$, where $\hbar$ is h divided by 2π, has a magnitude of about 200 MeV fm, where MeV stands for millions of electronvolts (room temperature corresponds to roughly 1/40 eV).

6. Frank Close, *Nuclear Physics: A Very Short Introduction*, Oxford University Press, 2015, p. 57.

7. There are also subtleties related to mirror symmetry, which go beyond the scope of this book. See Frank Close, *The New Cosmic Onion*, London, Taylor & Francis, 2007, p. 57.

8. Technically, in an antisymmetric quantum state.

9. In principle, the Higgs field could have given the photon a mass. However, the Higgs field in our universe happens to be blind to photons, leaving them massless. This aspect of the theory was first identified by Tom Kibble, 'Symmetry Breaking in Non-Abelian Gauge Theories', *Physical Review*, vol. 155, 1967, p. 1554. See Frank Close, *The Infinity Puzzle*, Oxford University Press, 2013, p. 171.

CHAPTER 5

1. Weinberg, *Dreams of a Final Theory*, p. 211.

2. As we saw in the last chapter, the magnitude of $\hbar c$ in units used by nuclear physicists is about 200 MeV fm, where 1 fm is 10^{-15} m and 1 MeV is a million electronvolts.

3. A kilowatt is a measure of power which is equivalent to 1,000 joules per second, so 1 kWh is equivalent to 3.6 million joules.
4. As to whether Einstein actually said that precise phrase, see Galina Weinstein, arxiv.org/abs/1310.1033.

CHAPTER 6

1. When energy density is fed into Einstein's equations and multiplied by other quantities that appear there, the dimension of the cosmological constant is found to be inversely proportional to area. The data correspond to the cosmological constant having a magnitude of 1 if the inverse area is equivalent to that of a square whose sides are of the order of 10 billion light years. This is, in effect, the distance that light from the Big Bang has travelled – in other words, the extent of the observable universe. However, the cosmological constant should have a magnitude of about 1, not when spread over the observable universe, but when concentrated on a square whose sides are the Planck length, about 10^{-35} metres.
2. I use 'string' theory when referring to the general ideas, and 'superstring' for when string theory is combined with supersymmetry.
3. Robert Laughlin, *A Different Universe*, New York, Basic Books, 2005, p. 212.
4. Comment to the author by George Ellis.
5. Comment to the author by George Ellis and Joe Silk.
6. Comment to the author by George Ellis.
7. Comment to author by Steven Weinberg.
8. As quoted in Frank Wilczek, *A Beautiful Question*, London, Allen Lane, 2015, p. 50.
9. Laughlin *A Different Universe*, p. 200.
10. As quoted by Nima Arkani-Hamed, www.sns.ias.edu/ckfinder/userfiles/files/daed_a_00161(3).pdf.

Further reading

Here I list books, articles and websites that I have found helpful.

THE FINAL THEORY

When he became the Lucasian Professor of Mathematics at Cambridge University in 1980, Stephen Hawking lectured an august assembly and put to them this question: 'Is the end in sight for theoretical physics?' His answer was that there was a fifty–fifty chance that by the end of the twentieth century there would be a fully unified theory. His lecture is reprinted in *Black Holes and Baby Universes and Other Essays* (London, Bantam Books, 1994). There has been considerable discussion of what he said and what it means, and argument about whether it is correct – see for example www.timeone.ca/hawking-end-physics/#sthash. JT6t9rIb.13Vw35KR.dpuf and affinemess.quora.com/On-Hawkings-G%C3%B6del-and-the-End-of-Physics. For an idea of the level of debate, see www.physicsforums.com/threads/hawkings-end-of-physics.583692/.

Lord Kelvin's speech was reported in the *Philosophical Magazine* in 1901 and may be read at blog.sciencenet.cn/home.php?mod=attachment&filename=19c%20clouds. pdf&id=54606. Michelson's remark about the sixth place of decimals appears to originate in a talk given at the University of Chicago in 1894, according to Steven Weinberg in *Dreams of a Final Theory* (New York, Vintage, 1994). Although Weinberg's book is over twenty years old, it is still the best survey of the history and philosophy of quests for a final theory. In it, he argues strongly that quantum mechanics will be part of any final theory. Paul Dirac's remark mentioned in Chapter 4 of the present book, that the 'underlying physical laws ... are completely known', is contained in P. A. M. Dirac, 'Quantum Mechanics of Many-Electron Systems', *Proceedings of the Royal Society A*, vol. 123, 1929, p. 714.

CLASSICAL MECHANICS

Isaac Newton set out his laws of mechanics and gravity in his book *Philosophiae Naturalis Principia Mathematica*; for a translation see *The Principia: The Authoritative Translation and Guide* (translated by I. Bernard Cohen and Anne Whitman, Berkeley, University of California Press, 2016). The beauty and power of the mathematics which enables calculation of orbits and many other applications are described in Ian Stewart's *Calculating the Cosmos* (London, Profile Books, 2016).

QUANTUM PHYSICS

A good place to start on the quest for enlightenment into the mysteries of the quantum universe is *In Search of*

Schrödinger's Cat (New York, Bantam Books, 1984) by John Gribbin, and *The Quantum Universe* by Tony Hey and Patrick Walters (Cambridge University Press, 1987). Hey and Walters's updated edition of their book, with much new material, is *The New Quantum Universe* (Cambridge University Press, 2003). The most complete history of the birth of quantum theory and the development of quantum mechanics is Abraham Pais's classic *Inward Bound* (Oxford University Press, 1986). This book also contains much mathematical commentary for those who wish to explore the depths of this subject. A more recent, highly readable and non-mathematical book on this area is Graham Farmelo's *The Strangest Man* (London, Faber & Faber, 2009). This biography of Paul Dirac gives a wonderful exposition of how he created the equation that now bears his name, together with his invention of quantum electrodynamics and its application.

THERMODYNAMICS

The emergence of the arrow of time has been covered in many books. *The Arrow of Time* by Peter Coveney and Roger Highfield (London, W. H. Allen, 1990) is an excellent place to delve deeper into this subject. In order to appreciate the conundrum of the entropy at the Big Bang, *From Eternity to Here* by Sean Carroll (New York, Dutton, 2010) is recommended; Carroll also discusses how the emergence of life on Earth is consistent with the laws of thermodynamics.

RELATIVITY

There are countless books on relativity at all levels from introductions to college texts. A non-mathematical account, highly illustrated, which also shows how relativity impacts our daily lives, is *Einstein's Mirror* by Tony Hey and Patrick Walters (Cambridge University Press, 1997). Einstein's life and work, together with some explanation of relativity, are in Abraham Pais's classic *Subtle Is the Lord* (Oxford University Press, 2005).

NUCLEAR AND PARTICLE PHYSICS

In *A Beautiful Question* (London, Allen Lane, 2015), Frank Wilczek gives an entertaining description of the present core theory of particles and forces, the standard model, with an emphasis on the role of symmetry. He gives a detailed introduction to the work of James Clerk Maxwell, and the diagrammatic summary in Figure 1 (see page 38) is based on his conception. Wilczek explains how Maxwell was inspired to modify Ampère's law to create equations with a mathematical symmetry. This book is also an excellent introduction to the ideas of coloured quarks and quantum chromodynamics, the area for which Wilczek won a Nobel Prize.

I have explained the foundations of nuclear physics and particle physics for non-experts in two pocket-sized books in Oxford University Press's 'Very Short Introduction' series: *Particle Physics – A Very Short Introduction* (2004) and *Nuclear Physics – A Very Short Introduction* (2015), and, in more detail, in *The New Cosmic Onion* (London, Taylor & Francis, 2007). I set out the history and development of modern particle theory in *The Infinity Puzzle* (Oxford University Press, 2011), which contains a description of renormalisation and the Higgs boson.

SUPERSYMMETRY AND SUPERSTRINGS

The basic ideas of supersymmetry are described in Gordon Kane's *Supersymmetry* (New York, Perseus Books, 2000). Making the case for superstring theory is Brian Greene's highly readable *The Elegant Universe* (New York, W. H. Norton, revised edition, 2003). Although superstrings are widely known, they are not the only theoretical attempt to build a quantum theory of gravity. To get a sense of the wider field, a good place to start is Peter Woit's highly critical survey of attempts to construct unified theories. The full title of the book says it all: *Not Even Wrong: The Failure of String Theory and the Continuing Challenge to Unify the Laws of Physics* (London, Jonathan Cape, 2006). Woit also questions whether string theory is really science, and describes some other attempts to create a quantum theory of gravity, such as 'loop quantum gravity'.

EMERGENCE

The idea that what at first appears fundamental has actually emerged from some more basic property is the theme of Robert Laughlin's *A Different Universe* (New York, Basic Books, 2005). The emergence of classical mechanics from quantum laws is a central theme of Laughlin's thesis. This picture of quantum mechanics, which originates with Richard Feynman, is also mentioned in John and Mary Gribbin's *Richard Feynman – A Life in Science* (New York, Dutton, 1997). This book introduces quantum electrodynamics and gives the historical background that led to Feynman's ideas. For an elegant exposition of the ideas of quantum electrodynamics by their creator, Richard Feynman's *QED – The Strange Theory of Light and Matter* (Princeton Science Library, 1985) is perfect.

Will space and time eventually emerge from some deeper theory? This is the question posed at the end of an excellent survey of the current state of theoretical physics and the quest for a theory of everything by Princeton theorist Nima Arkani-Hamed. His article, which inspired some of Chapter 7, is 'The Future of Fundamental Physics', published in *Daedalus*, the Journal of the American Academy of Arts and Sciences, in 2012. An electronic version can be found at www.sns.ias.edu/ckfinder/userfiles/files/daed_a_00161%283%29.pdf.

Acknowledgements

I am indebted to John Davey for suggesting this theme, and to John Woodruff and Paul Forty for helping bring it to fruition.

Index

INDEX

V

vacuum energy 109–111

vector fields 47, 105–106, 142

vectors
    electric and magnetic fields 105
    momentum 106

virtual particles 57, 106, 108

Vulcan (nonexistent planet) 97–8

W

W boson
    and the electroweak plasma 77
    emission of negatively-charged W⁻ 69
    exchange under the weak force 65
    mass of, and the Higgs field 74, 77–8
    as QFD force carrier 65–7, 105
    W⁺ and W⁻ 65, 69

Watt, James 24

the watt (unit) 27

wavefunctions 51

wavelength
    and the plasma limit 75
    proportional to particle momentum 50
    and temperature in the electromagnetic spectrum 82–3

weak force
    particles experiencing 65
    QED and 56
    QFD and 64

weightlessness 91–2

Weinberg, Stephen 79, 113, 128, 134

work, equivalence with heat 27

Y

year 1865 21, 28, 33–35

Z

Z boson 105, 115